Advanced Model-Based Charging Control for Lithium-Ion Batteries

Quan Ouyang · Jian Chen

Advanced Model-Based Charging Control for Lithium-Ion Batteries

華中科技大學出版社
Huazhong University of Science and Technology Press

Springer

Quan Ouyang
College of Automation Engineering
Nanjing University of Aeronautics
and Astronautics
Nanjing, Jiangsu, China

Jian Chen
School of Mechanical Engineering
Zhejiang University
Hangzhou, Zhejiang, China

ISBN 978-981-19-7058-0 ISBN 978-981-19-7059-7 (eBook)
https://doi.org/10.1007/978-981-19-7059-7

Jointly published with Huazhong University of Science and Technology Press
The print edition is not for sale in China (Mainland). Customers from China (Mainland) please order the print book from: Huazhong University of Science and Technology Press.

This Springer imprint is published by the registered company Springer Nature Singapore Pte Ltd.
The registered company address is: 152 Beach Road, #21-01/04 Gateway East, Singapore 189721, Singapore

Preface

Rechargeable lithium-ion batteries have been widely used for energy storage in numerous industries stretching from electric vehicles to microgrids due to their advantages of high energy density, long cycle life, and declining costs. Charging is an important process for lithium-ion batteries to replenish and store energy, and the quality of the charging strategy greatly affects the performance and lifetime of lithium-ion batteries. With accurate mathematical models to analyse and predict the changes of the battery's states during the charging process, advanced model-based charging strategies can provide excellent charging performance, such as delaying the degradation of battery life. Therefore, it is of great engineering and academic value to research advanced model-based lithium-ion battery charging control strategies.

Motivated by this, this book will introduce the state-of-the-art advanced model-based lithium-ion battery charging control technologies from the fundamental theories to practical designs and applications, especially in terms of battery modeling, state estimation, and optimal charging control. In addition, some other necessary design considerations, such as battery pack charging control with centralized and leader-followers structures, are also introduced to provide excellent solutions for improving the charging performance and extending the lifetime of the batteries/battery packs. The rich materials and knowledge presented in this book can give sufficient insight into the battery charging control technologies from the theoretical design to engineering applications. This brief is mainly divided into three parts and its organizational structure is as follows:

- The first part (Chaps. 1 to 3) of this book is devoted to providing an overview of the classifications and chemistry mechanisms of lithium-ion batteries. The fundamental conception of advanced model-based battery charging control and commonly utilized battery models are also described.
- The second part (Chaps. 4 and 5) aims to introduce two observers to provide the accurate estimated state of charge required in the model-based battery charging control strategies.
- The third part (Chap. 6) is concerned with a user-involved battery charging control strategy with economic cost optimization.

- The fourth part (Chaps. 8 to 10) firstly analyses the charging problem of battery packs and then introduce some advanced model-based charging control technologies for battery packs, including user-involved charging control strategies with centralized and leader-followers structures and fast charging control for battery packs.
- Finally, some future trends of battery charging management are mentioned in Chap. 11.

This work was supported by the National Natural Science Foundation of China (No. 61903189), the China Postdoctoral Science Foundation (No. 2020M681589), and the Key Research and Development Program of Zhejiang Province, China (No. 2021C01098).

Nanjing, China Quan Ouyang
Hangzhou, China Jian Chen
May 2022

Contents

Acronyms

AC	Alternating-Current
BMS	Battery Management System
CC	Constant Current
CC-CV	Constant Current-Constant Voltage
CV	Constant Voltage
DC	Direct-Current
DICM	Discontinuous Inductor Current Mode
EKF	Extended Kalman Filter
ESI	Electrochemical Impedance Spectroscopy
EV	Electric Vehicle
FUDS	Federal Urban Driving Schedule
GA	Genetic Algorithm
GPIC	General Purpose Inverter Controller
IDA-PBC	Interconnection and Damping Assignment-Passivity-Based Controller
LMI	Linear Matrix Inequality
LS	Least Squares
MPC	Model Predictive Control
Ni-Cd	Nickel-Cadmium
Ni-MH	Nickel-Metal Hydride
OCV	Open Circuit Voltage
P2D	Pseudo-Two-Dimensional
PID	Proportional-Integral-Derivative
PNGV	Partnership for New Generation of Vehicles
PSO	Particle Swarm Optimization
PWM	Pulse Width Modulation
RBF	Radial Basis Function
RLS	Recursive Least Squares
RMS	Root Mean Square

SEI	Solid Electrolyte Interphase
SOC	State of Charge
SOH	State of Health
TOU	Time-of-Use

Chapter 1
Introduction

1.1 Brief Introduction of Lithium-Ion Batteries

Under the background of increasing energy demand and serious environmental crisis (as illustrated in Fig. 1.1), the world is shifting from fossil fuels to renewable energy sources. The renewable energy industries, such as wind energy and photovoltaic, have ushered in opportunities for leapfrog development. Renewable energy is green and low-carbon, which can effectively contribute to energy structure optimization, ecological environment protection, and sustainable economic and social development. But it brings an urgent problem, i.e., how to store and use the generated renewable energy efficiently. Electrochemical energy storage technologies represented by rechargeable batteries have become the most popular energy storage solution [1] since they can directly store and release electrical energy without being restricted by the geographical and terrain environment.

1.1.1 *Comparison with Other Commonly Used Batteries*

Currently, the widely used batteries mainly include lead-acid, nickel-cadmium (Ni-Cd), nickel-metal hydride (Ni-MH), and lithium-ion batteries [2]. Lead-acid batteries are widely used in automobiles and electric motorcycles because of their mature production technology, high safety, and low price. However, they have a low gravimetric specific power and energy density. The Ni-Cd batteries are ideal direct-current (DC) power suppliers since they have good charge-discharge rate performance. However, they have defects of memory effect and low nominal voltage. Moreover, due to the environmental pollution caused by the heavy metal cadmium, the Ni-Cd batteries are being phased out. Ni-MH batteries have the advantages of a large charge/discharge rate and little memory effect. However, they have a low nominal voltage and are not suitable to be used in parallel. Lithium-ion batteries have no memory effect and no

Q. Ouyang and J. Chen, *Advanced Model-Based Charging Control for Lithium-Ion Batteries*, https://doi.org/10.1007/978-981-19-7059-7_1

Fig. 1.1 Energy and environmental crisis

Table 1.1 Performance comparison of commonly used batteries [3]

	Lead-acid	Ni-Cd	Ni-MH	Lithium-ion
Nominal voltage (V)	2	1.2	1.2	3.7
Gravimetric specific power (Wh/kg)	30–50	40–50	50–70	120–140
Energy density (Wh/L)	60–100	80–100	100–140	240–280
Cycle life (times)	400–600	800–2000	800–2000	1200
Cost (dollars/kWh)	120–150	300–350	150–200	150–180

pollution. Their gravimetric specific power and energy density are much higher than other types of batteries.

The performance comparison results of the above-mentioned batteries are illustrated in Table 1.1 [3]. The nominal voltage of a single lithium-ion battery is about 3.7 V, which is equal to the voltage of 3 serial connected Ni-Cd or Ni-MH batteries. The gravimetric specific power and energy density of the lithium-ion batteries are approximately 1.5–3 times that of Ni-Cd batteries. It demonstrates the excellent performance of lithium-ion batteries.

1.1.2 Applications of Lithium-Ion Batteries

Due to the advantages of high power/energy density, long cycle life, and low self-discharge rate, the lithium-ion batteries have been widely used in numerous industrial

(a) Tesla EV

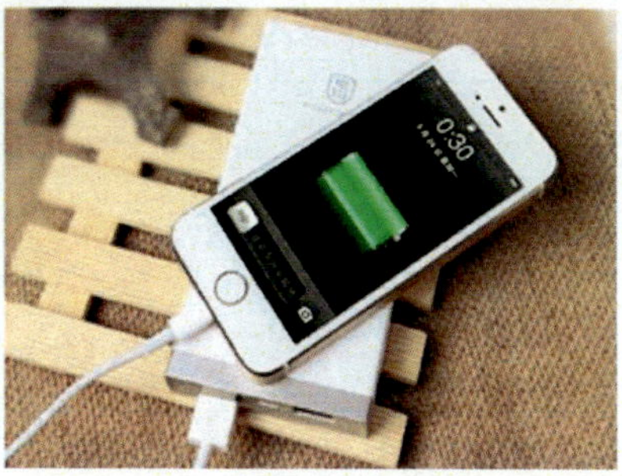

(b) Iphone

(c) Large-scale energy storage system

(d) Unmanned aerial vehicle

Fig. 1.2 Applications of lithium-ion batteries

applications stretching from electric vehicles (EVs) and portable electronic products to large-scale energy storage systems and unmanned aerial vehicles [4], as illustrated in Fig. 1.2. From the report by Foresight Industry Research Institute [5], the global lithium-ion battery demand exceeded 279 GWh in 2020, and it is estimated that the demand will reach as high as 1223 GWh by 2025 (as shown in Fig. 1.3). The global market structure of lithium-ion batteries is shown in Fig. 1.4, where the lithium-ion batteries have occupied the dominance of the energy storage market.

1.2 Format Comparison of Lithium-Ion Batteries

Due to different manufacturing and packaging methods, there exist four available basic formats of lithium-ion batteries: cylindrical, prismatic, pouch, and special-shaped, as shown in Fig. 1.5.

- Cylindrical lithium-ion batteries have a long history of development, where the earliest one was the 18650 battery invented by SONY in 1992. Its production technology is mature and highly automated, which contributes to a high yield

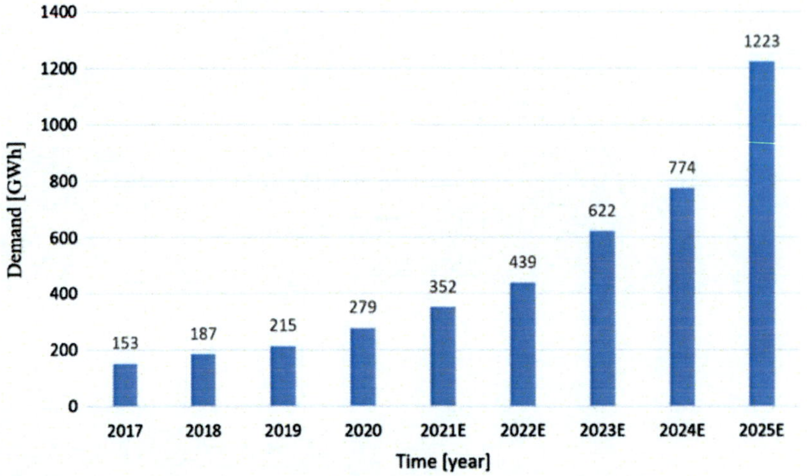

Fig. 1.3 Global lithium-ion battery demand [5]

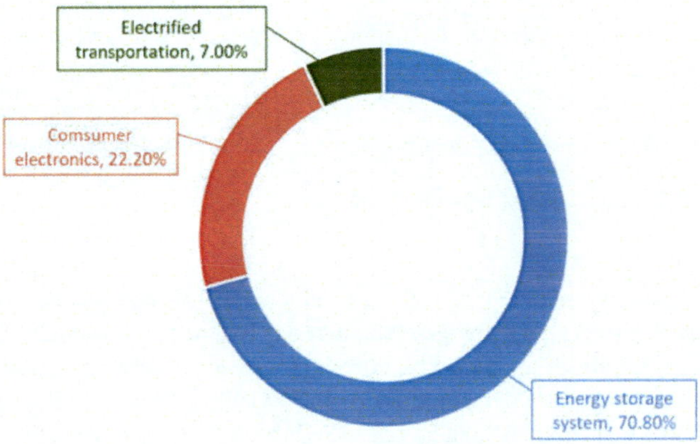

Fig. 1.4 Global market structure of lithium-ion batteries [6]

rate and good product consistency. Since the heat dissipation area of the cylindrical batteries is larger than the batteries with other formats, they have better heat dissipation performance. Moreover, the cylindrical shape is conducive to the dispersion of force, thereby reducing the risk of thermal runaway of the battery caused by mechanical collision. However, cylindrical lithium-ion batteries have the disadvantages of low capacity and high grouping complexity.

- The production technology of prismatic lithium-ion batteries based on winding or lamination is mature and reliable. They have the advantages of low price and customizable size and have been widely used in early EVs for a long time. However, since the prismatic batteries in the battery pack are tightly packaged together with

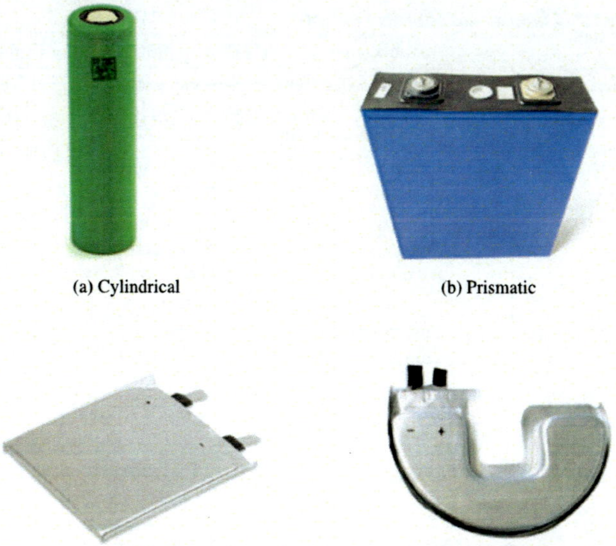

(a) Cylindrical (b) Prismatic

(c) Pouch (d) Special-shaped

Fig. 1.5 Different formats of lithium-ion batteries

only leaving a small gap, it results in poor heat dissipation performance and even causes an explosion when thermal runaway occurs.

- The biggest difference between the pouch battery and other lithium-ion batteries is that it has aluminum-plastic film packaging. When the temperature is too high or even thermal runaway occurs, the pouch battery will produce bulging to reduce the risk of explosion, which can bring higher safety. The pouch batteries also have the advantages of light weight, high energy density, and flexible design. But there are some urgent challenges for pouch batteries, such as poor consistency, risk of electrolyte leakage, and high cost.
- The special-shaped batteries can be manufactured into various shapes according to different demands. They can sufficiently utilize the irregular available space of the product and are widely used in smart wearable devices, such as smart bracelets. Their thickness can be as low as 0.45 mm, which caters to the needs of the wearable market. The special-shaped batteries also have high mechanical stability, durability, and safety. But they have the problem of high cost.

From the above analysis, the lithium-ion batteries with different formats have their unique advantages and suitable application scenarios, where their comparison results are summarized in Table 1.2. Note that the above-mentioned lithium-ion batteries have the same chemistry mechanism. It means that the charging control strategies mentioned in this book can be applied to all these batteries.

Table 1.2 Comparisons of battery formats

	Advantages	Disadvantages	Suitable applications
Cylindrical	Mature production technique Good heat dissipation	Low capacity High grouping complexity	EVs Backup energy
Prismatic	Low price Customizable size	Poor heat dissipation in the pack	EVs
Pouch	High energy density Light weight High safety	Poor consistency Risk of electrolyte leakage High cost	Consumer electronics such as mobile phones
Special-shaped	Customizable shape Thin and light	High cost	Smart wearable devices

1.3 Electrochemical Mechanism of Lithium-Ion Batteries

1.3.1 Composition of Lithium-Ion Batteries

The lithium-ion battery is mainly composed of positive and negative electrodes, electrolyte, separator, etc., relying on lithium ions to move between the positive and negative electrodes with electrolyte as the medium to realize the charge and discharge work [7]. Lithium-ion batteries usually use an electrolyte solution made of LiPF6 dispersed in alkyl carbonates. These include ethylene carbonate and another solvent comparable to ethyl methyl carbonate. Electrolytes are unstable at low potentials, so when an electrolyte decreases on graphite surfaces, it forms a solid electrolyte interphase (SEI). The SEI layer possesses both an electronically insulating and an ionically conductive property, causing reaction rates to decrease. As a result, a lithium-ion battery's longer life span is dependent on the stability of the SEI layer. Hence it is essential that this layer be maintained over time. To achieve a more stable SEI layer, an electrolyte additive, such as vinylene carbonate or fluoroethylene carbonate, or mixtures of electrolytes can be added to the electrolyte. Although electrolyte additives are responsible for stabilizing the SEI layer, there is a possibility they could also reduce cathode surface oxidation at higher potentials through the electrolyte.

Usually, the graphite is utilized as the negative electrode of the battery. Intercalation is the process of inserting lithium ions between graphite sheets, and it is the result of an insertion mechanism of ions into the graphite structure. An atom of Li can intercalate between six carbon atoms of graphite at room temperature, resulting in a capacity of 372 mAh/g. However, in practice graphite can't surpass its theoretical capacity. A battery's capacity of graphite changes after early charge-discharge cycles. This can happen when irreversible consumption of active lithium creates the SEI layer, which changes the battery's capacity.

The determination of a suitable material for anodes or cathodes in lithium-ion batteries is dependent on a combination of active materials, a conductive agent, as well

Table 1.3 Cathode materials overview and evaluation [8]

	$LiCoO_2$	$LiNiO_2$	$LiMn_2O_4$	$Li(Ni_xCo_yMn_z)O_2$	$LiFePO_4$
Voltage Li/Li$^+$	3.9 mV	3.8 mV	4.0 mV	3.8 mV - 4.0 mV	3.4 mV
Capacity	150 mAh/g	170 mAh/g	120 mAh/g	130–180 mAh/g	180 mAh/g
Safety	–	–	+	0	++
Stability	–	–	0	0	++
Price	–	–	+	0	+

as a binder. In the electrode composite, the percentage of each component determines the performance of the final electrode. The resulting combination is usually referred to as slurry and is coated on a suitable current collector. For academic research environments, the coating is typically applied by hand or by means of an automated doctor-blade film coater. As a step toward improving conductivity and density, the electrodes must be compressed before they can be used. A compressed electrode has a reduced surface area, resulting in fewer side reactions.

Positive electrodes are commonly composed of transition metal oxides in which Li-ions can reversibly insert themselves within the structure. Subsequently, cathode-induced thermodynamic reactions allow lithium ions to be sucked into the cathode during discharge. According to the chemical materials of the positive electrode, lithium-ion batteries can be named as $LiCoO_2$, $LiMnO_2$, $LiMn_2O_4$, $LiFePO_4$, and so on. Table 1.3 provides a brief overview of the various cathode materials [8].

1.3.2 Charging-Discharging Mechanism

Although the nominal voltage, capacity, and gravimetric specific power vary with these batteries, their chemistry mechanism is the same [9], as shown in Fig. 1.6.

- In the charging mode, lithium ions are generated through the chemical reaction in the positive electrode of the lithium-ion battery. And then, these lithium ions pass through the separator through the electrolyte to reach the negative electrode of the battery. Since the negative electrode is made of carbon material with a layered microporous structure, these lithium ions are inserted and stored in the micropores of the carbon layer. The energy stored in the lithium-ion battery increases with the continuous insertion of the lithium ions embedded in the negative electrode.
- In the discharging mode, lithium ions embedded in the carbon layer are released from the negative electrode, pass through the separator and return to the positive electrode of the lithium-ion battery. The energy stored in the lithium-ion battery decreases with the continuous return of the lithium ions to the positive electrode.

Take the $LiCoO_2$ battery as an example. When it is in the charging mode, its chemistry reaction is as follows:

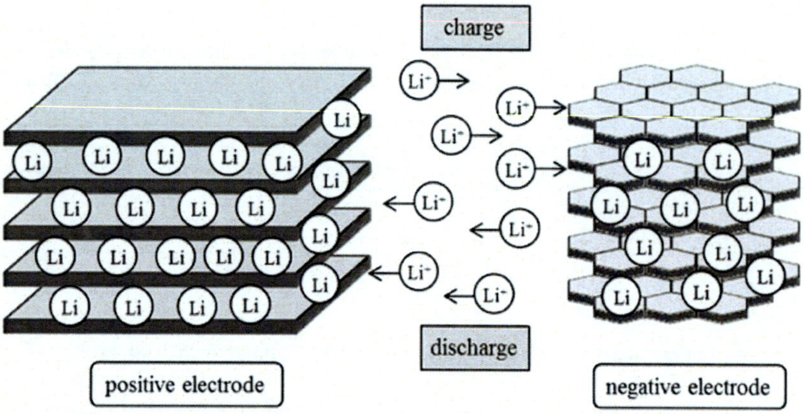

Fig. 1.6 Chemistry mechanism of lithium-ion batteries [10]

$$\begin{aligned} LiCoO_2 &\rightarrow Li_{1-x}CoO_2 + xLi^+ + xe^- \quad \text{(positive electode)} \\ 6C + xLi^+ + xe^- &\rightarrow Li_xC_6 \quad \text{(negative electode)}. \end{aligned} \tag{1.1}$$

The overall reaction is as

$$LiCoO_2 + 6C \rightarrow Li_{1-x}CoO_2 + Li_xC_6. \tag{1.2}$$

Similarly, when the lithium-ion battery is in the discharging mode, its overall reaction is

$$Li_{1-x}CoO_2 + Li_xC_6 \rightarrow LiCoO_2 + 6C. \tag{1.3}$$

In fact, the electrochemical reaction inside a lithium-ion battery is highly complicated. The following operations should be avoided for better battery use.

- Overcharging. When the lithium-ion battery is overcharged, lithium ions will be excessively released, which brings significant damage to the structure of the positive electrode. At the same time, there are too many lithium ions that fail to be inserted into the negative electrode, causing lithium precipitation on the negative electrode surface, thus resulting in irreversible damage to the lithium-ion battery.
- Over-voltage. The high voltage caused by over-voltage charging will make the electrolyte decompose and generate a large amount of gas, thus causing the lithium-ion battery to swell and even explode in severe cases.
- Over-discharging. When the lithium-ion battery is over-discharged, too many lithium ions migrate out of the negative electrode, leading to the collapse of the negative plate layer.

(a) Pouch battery (b) Phone battery

Fig. 1.7 Bulge of lithium-ion batteries

- High temperature. The high temperature will cause the shrink or rupture of the internal separator of the lithium-ion battery. The thermal runaway can result in fire and explosion.
- High current. When the charging/discharging current is excessively high, lithium ions will accumulate on the surface of the negative electrode, which will cause the precipitation of lithium crystals. It will bring irreversible damage to the negative electrode and sharply reduce the lifetime of the lithium-ion battery.

1.4 Motivation of Advanced Model-Based Battery Charging Control

Charging is a crucial process for lithium-ion batteries to replenish energy, and battery charging management plays a vital role in the optimization of the operation and life extension of lithium-ion batteries. Charging control is one of the most important functions in the battery management system (BMS). Due to the complex electrochemical reaction, lithium-ion batteries are sensitive to the charging protocol. Improper charging behaviors, such as overcharging or charging with excessive current, can cause lithium-ion precipitation and crystallization inside the battery, internal pressure increase, and temperature rise, which leads to the rapid capacity degradation and bulge of lithium-ion batteries, and even fire or explosion in severe cases (as shown in Figs. 1.7 and 1.8). To ensure the safety and high performance of lithium-ion batteries, much attention must be paid to charging management to avoid improper operations such as overcharging, over-voltage charging, and high current charging. On the other hand, a low charging speed will cause inconvenience in the battery use and eventually impair the consumer satisfaction level. This hence calls for effective charging strategies. A wide variety of battery charging methods that have been published recently fall into two basic categories: non-model-based and model-based techniques.

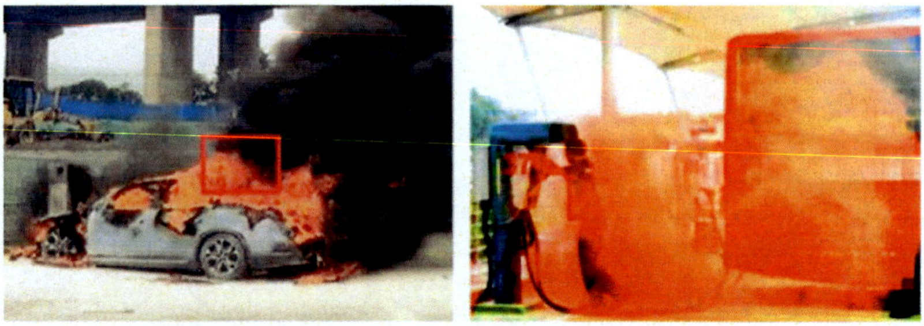

(a) Electric vehicle (b) Electric bus

Fig. 1.8 Fire caused by improper charging of lithium-ion batteries

1.4.1 Non-model-based Charging Control

Numerous non-model-based charging control techniques have been designed. The popular conventional charging ways are the constant voltage (CV) method and the constant current (CC) strategy, which applies a constant voltage or forces a constant current flow through the battery [11]. The CC charging method is a simple way to charge batteries steadily at a low current until they are fully charged. When the CC charging time reaches a predetermined level, the charging stops. In CC charging, a battery's behavior is highly influenced by the charging current. The CV charging is another widely used charging method in which a constant voltage is used to charge the batteries, which can avoid overvoltage and irreversible side reactions of the battery. Through easy for implementation, they can lead to serious detrimental effects for the battery. One improvement is the constant current-constant voltage (CC-CV) strategy [12], as shown in Fig. 1.9, where the battery is charged with a constant current until its terminal voltage reaches the pre-set threshold and then continues to be charged with a constant voltage. The charging process is terminated when the charging current decreases gradually into a small enough value. Another commonly used method is the pulse charging strategy, whose current profile is based on pulses. There is a short rest period between two consecutive pulses that makes the lithium-ion battery charge in a short standstill or reverse discharge state between consecutive pulses, which effectively alleviates the polarization effect, thereby slowing down the capacity fade [13].

To improve the charging speed, the multi-step constant-current charging strategy is proposed in [14], which divides the entire charging process into several constant-current charging stages with a decreasing trend of the charging current. When the voltage of the lithium-ion battery reaches the limit value, the charging pattern automatically switches to the next constant current stage until all preset charging processes are completed.

However, these non-model-based charging strategies are empirical by design without considering the battery's dynamics and will suffer performance loss given a poor selection of charging parameters, e.g., the magnitude of the constant current. More-

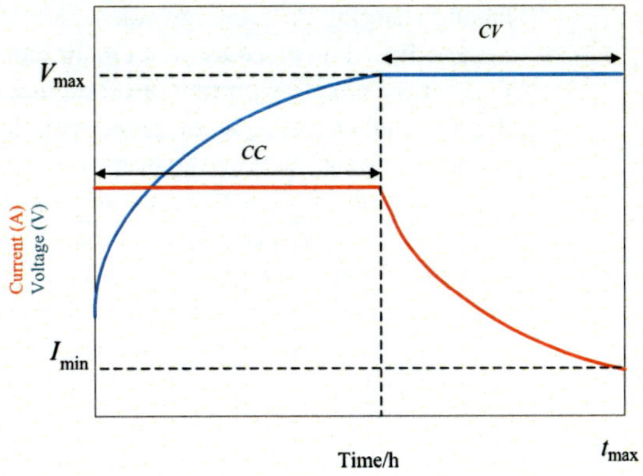

Fig. 1.9 Charging profile for CC-CV

over, the fixed preset charging parameters may bring damage to the battery due to the mismatch of charging parameters and battery characteristics, causing a decrease in battery performance and lifetime.

1.4.2 Model-Based Charging Control

To remedy the above mentioned deficiencies of the non-model-based control strategies, accurate mathematical models that capture the lithium-ion battery's dynamics can be utilized in the charging control algorithm design, which can help analyse and predict the changes of the battery's states and parameters during the charging process, thus improving the charging efficiency and delaying the degradation of battery life. Therefore, it is of great engineering and academic value to research advanced model-based lithium-ion battery charging control strategies.

The existing model-based charging control strategies can be mainly divided into two categories:

(1) **Offline optimization of charging current based on battery model**. A particle swarm optimization based fuzzy-controlled searching strategy is developed in [15] to determine the optimal multi-stage charging pattern that can deliver the most discharged capacity within the shortest charging time. In [16], an optimal control theory-based charging strategy is proposed with aiming to obtain an optimal current profile to minimize the charging time, energy loss, and temperature rise. In [17], a fast charging method is proposed that performs a multi-objective optimization between charging time and temperature rise, and a genetic algorithm is employed to search for its optimal charging current trajectories. Based on the coupled thermoelectric model, a heuristic algorithm is designed in [18] to obtain the optimal constant current for the battery, which can achieve a balance

among charging time, energy conversion efficiency, and temperature rise of the battery. Based on an equivalent circuit model, an optimal pattern is designed in [19] for the charging current design optimization, which can reduce the charging time and energy loss compared with the CC-CV strategy. A high-efficiency adaptive current charging algorithm is developed in [20] with considering the variation of the battery's internal resistance that can decrease the charging loss compared with the conventional constant current charging method. The Taguchi method is developed in [21] to search the optimal charging pattern for a five-stage constant-current charging strategy, where the charging capacity, energy conversion efficiency, and charging time are considered simultaneously. Based on the electrothermal-aging dynamics of the battery, an optimal multi-stage charging method is proposed in [22] with a balance between charging time and capacity fade achieved. A health-aware multi-objective optimal charging method is developed in [23] that could simultaneously increase the charging speed and relieve the degradation of the battery. In [24], an optimal charging strategy is designed for aged batteries with considering the charging time and temperature rise into consideration. A data-driven Bayesian optimization framework is proposed in [25] to achieve fast charging while satisfying battery degradation-related constraints.

(2) **Real-time adjustment of charging patterns based on battery's state feedback**. In [26], a fast charging strategy based on model predictive control is proposed with the aim of simultaneously reducing the charge duration and the increase in temperature of the battery. A charging control algorithm based on the reference governor is designed in [27], which can improve the charging capacity and charging speed while ensuring the safety constraints of the battery during the charging process. Based on a coupled electrothermal model, a linear-time-varying model predictive control algorithm is formulated in [28] to enable the fast battery charging, and the simulation results demonstrate that a balance of the charging time and temperature rise can be obtained through using this approach. A constrained generalized predictive control strategy is proposed in [29], which can enable fast charging while limiting temperature rise of the battery. C. Zou [30] proposed a model predictive control method to online tune the lithium-ion battery charging process to achieve the best trade-off between battery charging time and health protection. In [31], an explicit model predictive control (MPC) algorithm is developed for battery charging control, which can reduce the computational complexity of the traditional MPC method by pre-computing explicit solutions as piecewise functions. In [32], an online optimal fast charging strategy is proposed based on a reduced electrochemical model. A novel distributed electrochemical state estimation and fast-charging method for lithium-ion batteries using a sophisticated physics-based model are presented in [33].

References

1. A. Kaabeche, Y. Bakelli, "Renewable hybrid system size optimization considering various electrochemical energy storage technologies," *Energy Conversion and Management*, vol. 103, pp. 162–175, 2019.
2. Q. Ouyang, *Research on Key Technologies of Lithium-Ion Battery Management System for Electric Vehicles*, Ph.D. thesis, Zhejiang University, China, 2018.
3. S. Zhang, *Research on Equalization Algorithm Based on LiFePO4 Cell State of Charge*, Ph.D. thesis, Shanghai Jiaotong University, China, 2015.
4. X. Hu, J. Jiang, D. Cao, and B. Egardt, "Battery health prognosis for electric vehicles using sample entropy and sparse Bayesian predictive modeling," *IEEE Transactions on Industrial Electronics*, vol. 63, no. 4, pp. 2645–2656, 2016.
5. Foresight Industry Research Institute, *Report of Market Demand Forecast and Investment Strategy Planning on China Li-ion Power Battery Industry (2021–2026)*, Foresight Industry Research Institute, China, 2021.
6. Foresight Industry Research Institute, *Report of Market Prospective and Investment Strategy Planning on China Energy Storage Battery Industry (2021–2026)*, Foresight Industry Research Institute, China, 2021.
7. M. Yoshio, J. B. Ralph, and K. Akiya, *Lithium-Ion Batteries*, New York: Springer, 2009.
8. H. M. Heyn, *Adaptive State Estimation of Lithium-Ion Batteries in electric vehicles*, Master thesis, University of Gothenburg, Sweden, 2013.
9. D. Andrea, *Battery Management Systems for Large Lithium-Ion Battery Packs*, London: Artech House, 2010.
10. J. Garche, C. Dyer, P. Moseley, Z. Ogumi, D. Rand, and B. Scrosati, *Encyclopedia of Electrochemical Power Sources*, Elsevier, Amsterdam, 2013.
11. R. Garcia-Valle, J. A. P. Lopes, *Electric Vehicle Integration into Modern Power Networks*, New York NY, USA: Springer, 2012.
12. S. Zhang, K. Xu, and T. Jow, "Study of the charging process of a LiCoO2-based li-ion battery," *Journal of Power Sources*, vol. 160, no. 2, pp. 1349–1354, 2006.
13. A. Aryanfar, D. J. Brooks, B. V. Merinov, W. A. Goddard, A. J. Colussi, and M. R. Hoffmann, "Dynamics of lithium dendrite growth and inhibition: Pulse charging experiments and monte carlo calculations," *Journal of Physical Chemistry Letters*, vol. 5, no. 10, pp. 1721–1726, 2014.
14. T. Ikeya, N. Sawada, S. Takagi, J. Murakami, and K. Kobayashi, "Multi-step constant-current charging method for electric vehicle, valve-regulated, lead/acid batteries during night time for load-levelling," *Journal of Power Sources*, vol. 75, no. 1, pp. 101–107, 1998.
15. S. Wang, Y. Liu, "A PSO-based fuzzy-controlled searching for the optimal charge pattern of li-ion batteries," *IEEE Transactions on Industrial Electronics*, vol. 62, no. 5, pp. 2983–2993, 2015.
16. A. Abdollahi, X. Han, G. Avvari, N. Raghunathan, B. Balasingam, K. Pattipati, and Y. Bar-Shalom, "Optimal battery charging, part I: Minimizing time-to-charge, energy loss, and temperature rise for OCV-resistance battery model," *Journal of Power Sources*, vol. 303, pp. 388–398, 2016.
17. C. Zhang, J. Jiang, Y. Gao, W. Zhang, Q. Liu, and X. Hu, "Charging optimization in lithium-ion batteries based on temperature rise and charge time," *Applied Energy*, vol. 194, pp. 569–577, 2017.
18. K. Liu, K. Li, Z. Yang, and J. Deng, "An advanced lithium-ion battery optimal charging strategy based on a coupled thermoelectric model," *Electrochimica Acta*, vol. 225, pp. 330–344, 2017.
19. A. B. Khan, W. Choi, "Optimal charge pattern for the high-performance multistage constant current charge method for the li-ion batteries," *IEEE Transactions on Energy Conversion*, vol. 33, no. 3, pp. 1132–1140, 2018.

20. J. Ahn, B. K. Lee, "High-efficiency adaptive-current charging strategy for electric vehicles considering variation of internal resistance of lithium-ion battery," *IEEE Transactions on Power Electronics*, vol. 34, no. 4, pp. 3041–3052, 2019.
21. L. Jiang, Y. Li, Y. Huang, J. Yu, X. Qiao, Y. Wang, C. Huang, and Y. Cao, "Optimization of multi-stage constant current charging pattern based on Taguchi method for li-ion battery," *Applied Energy*, vol. 259, pp. 114–148, 2020.
22. X. Hu, Y. Zheng, X. Lin, and Y. Xie, "Optimal multistage charging of nca/graphite lithium-ion batteries based on electrothermal-aging dynamics," *IEEE Transactions on Transportation Electrification*, vol. 6, no. 2, pp. 427–438, 2020.
23. Y. Gao, X. Zhang, B. Guo, C. Zhu, J. Wiedemann, L. Wang, and J. Cao, "Health-aware multiobjective optimal charging strategy with coupled electrochemical-thermal-aging model for lithium-ion battery," *IEEE Transactions on Industrial Informatics*, vol. 16, no. 5, pp. 3417–3429, 2020.
24. J. Sun, Q. Ma, C. Tang, T. Wang, T. Jiang, and Y. Tang, "Research on optimization of charging strategy control for aged batteries," *IEEE Transactions on Vehicular Technology*, vol. 69, no. 12, pp. 14141–14149, 2020.
25. B. B. Jiang, M. D. Berliner, K. Lai, P. A. Asinger, H. B. Zhao, P. K. Herring, M. Z. Bazant, and R. D. Braatz, "Fast charging design for lithium-ion batteries via Bayesian optimization," *Applied Energy*, vol. 307, pp. 118–244, 2022.
26. J. Yan, G. Xu, H. Qian, Y. Xu, and Z. Song, "Model predictive control based fast charging for vehicular batteries," *Energies*, vol. 4, no. 8, pp. 1178–1196, 2011.
27. H. Perez, N. Shahmohammadhamedani, and S. Moura, "Enhanced performance of li-ion batteries via modified reference governors and electrochemical models," *IEEE/ASME Transactions on Mechatronics*, vol. 20, no. 4, pp. 1511–1520, 2015.
28. C. Zou, X. Hu, Z. Wei, and X. Tang, "Electrothermal dynamics-conscious lithium-ion battery cell-level charging management via state-monitored predictive control," *Energy*, vol. 141, pp. 250–259, 2017.
29. K. Liu, K. Li, and C. Zhang, "Constrained generalized predictive control of battery charging process based on a coupled thermoelectric model," *Journal of Power Sources*, vol. 347, pp. 145–158, 2017.
30. C. Zou, C. Manzie, and D. Nei, "Model predictive control for lithium-ion battery optimal charging", *IEEE/ASME Transactions on Mechatronics*, vol. 23, no. 2, pp. 947–957, 2018.
31. N. Tian, H. Fang, and Y. Wang, "Real-time optimal charging for lithiumion batteries via explicit model predictive control," in *International Symposium on Industrial Electronics*, 2019, pp. 2001–2006.
32. Y. Yin, S. Y. Choe, "Actively temperature controlled health-aware fast charging method for lithium-ion battery using nonlinear model predictive control," *Applied Energy*, vol. 271, p. 115232, 2020.
33. Y. Li, D. M. Vilathgamuwa, E. Wikner, Z. Wei, X. Zhang, T. Thiringer, T. Wik, and C. Zou, "Electrochemical model-based fast charging: Physical constraint-triggered PI control," *IEEE Transactions on Energy Conversion*, vol. 36, no. 4, pp. 3208–3220, 2021.

Chapter 2
Lithium-Ion Battery Charging Technologies: Fundamental Concepts

2.1 Definitions Related to Battery Charging

2.1.1 Basic Performance Parameters

2.1.1.1 Rated Capacity

The rated capacity of the battery refers to the minimum amount of electricity that the battery can discharge under a certain condition (with a certain temperature, current rate, and a cut-off voltage), whose unit is ampere-hours. The theoretical rated capacity of the battery can be calculated according to the amount of electrode active material in the battery electrochemical reaction formula and the electrochemical equivalent of the active material obtained by Faraday's law. However, the rated capacity of the battery under actual operating conditions is affected by many factors, such as the ambient temperature, current, etc. The actual rated capacity of the battery is often lower than the theoretical one due to the side reactions that may occur in the battery and the special needs of the design. Usually, the rated capacity of the battery is marked on the surface of the battery or on the manual. A lithium-ion battery with 5000 mAh manufactured by Panasonic is as shown in Fig. 2.1.

2.1.1.2 Open Circuit Voltage

The open circuit voltage (OCV) of the battery is the measured positive and negative potential difference of the battery when the battery is in the standby mode, i.e., the battery's current is zero. The parameters related to the OCV of the battery include the electrode material, the composition and concentration of the electrolyte, the temperature, and the state of the electrode interface. Note that there is an identified relationship between the OCV and the SOC of the battery. Usually, the OCV of the

Q. Ouyang and J. Chen, *Advanced Model-Based Charging Control for Lithium-Ion Batteries*, https://doi.org/10.1007/978-981-19-7059-7_2

Fig. 2.1 Panasonic battery

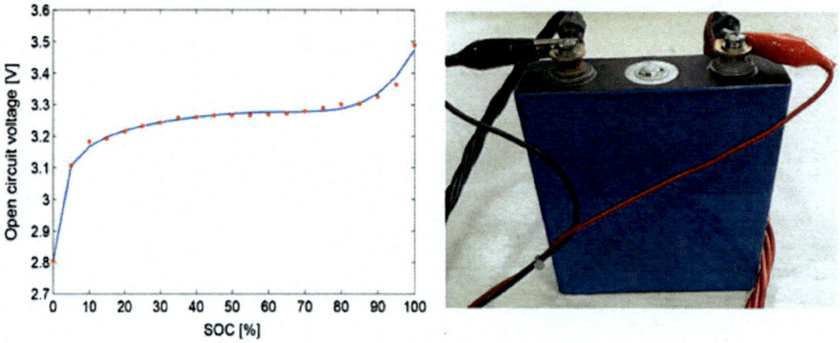

Fig. 2.2 OCV curves of an IFP36130155-36Ah lithium iron phosphate battery

battery is measured every 5% SOC, and the mapping from SOC to OCV can be obtained through the interpolation method. As an example, the OCV curve of an IFP36130155-36Ah lithium iron phosphate battery is illustrated in Fig. 2.2 [1].

2.1.1.3 Terminal Voltage

The terminal voltage, also known as the working voltage, refers to the potential difference between the positive and negative electrodes of the battery. The battery's terminal voltage V_B can be expressed as

$$V_B = V_{OC} - R_{in} I_B \tag{2.1}$$

where R_{in} denotes the battery's internal resistance, V_{OC} is the battery's OCV, and I_B the current, which is positive/negative when the battery is in the discharging/charging mode. From (2.1), it observes that there exists the ohmic voltage drop because of the internal resistance in the discharging state, which results in the battery's terminal voltage being lower than its OCV. The terminal voltage is higher than the OCV, when the battery is charged. Note that due to the potential limitations of cathode and anode electrodes, a single lithium-ion battery voltage is usually limited in the range of 2.5–4.2 V. The batteries can be connected in series and parallel to provide the needed high voltage in practice.

2.1.1.4 C-Rate

The C-rate reflects the relationship between the current and rated capacity and defines how fast or slow the cycle is performed. In this regard, 1 C-rate indicates that the specified current is capable of charging or discharging the battery within 1 h. 2 C-rate signifies that the specified current will last for an average of 30 min for each charge or discharge. Similarly, 1/2 C-rate indicates the battery can be fully charged or discharged in 2 h. As an example, for the cylindrical 18,650 battery with the rated capacity of 2600 mAh, 1 C-rate is 2.6 A. Usually, lithium-ion batteries with high maximum charging/discharging current rates requires higher production processes and are more expensive.

2.1.1.5 Self-discharge Rate

The battery self-discharge refers to the phenomenon of voltage drop and capacity loss after the battery is left standby for a period of time. Theoretically, the self-discharge of the battery can be zero, but in the practical applications, the self-discharge behavior of the battery is unavoidable due to the capacity loss of the spontaneous chemical reaction inside the battery, the micro-short-circuit loss inside the battery, and the defect of the diaphragm. The self-discharge process occurs inside the lithium-ion battery, which is related to the manufacturing process and changes with the ambient temperature, battery life, state of charge, discharge capacity, and number of cycles. Usually, the lithium-ion battery is left open-circuited for 7 days or 28 days, and the self-discharge rate is characterized by the rate of voltage reduction during storage. The self-discharge rate of lithium-ion batteries is about 1-5% per month.

2.1.1.6 Cycling Life

One cycling of the battery includes a complete charging and discharging process, and the cycling life is defined as the cycling number that the battery can perform while maintaining a certain capacity. The battery's capacity will slowly decay due to the reduction of lithium ions involved in energy transfer caused by the precipitation of metal lithium, the decomposition of positive electrode materials, the SEI film on the electrode surface, the loss of electrolytes, the aging and decay of battery materials. The battery's capacity decay is irreversible and accumulates over multiple cycles, adversely affecting battery performance. Although the lithium-ion battery has a longer cycle life compared with other types of batteries, improper charging behavior can lead to a significant reduction in battery cycling life. Taking the battery 05 in the NASA dataset as an example, the relationship between the capacity and cycling number is illustrated in Fig. 2.3 [2].

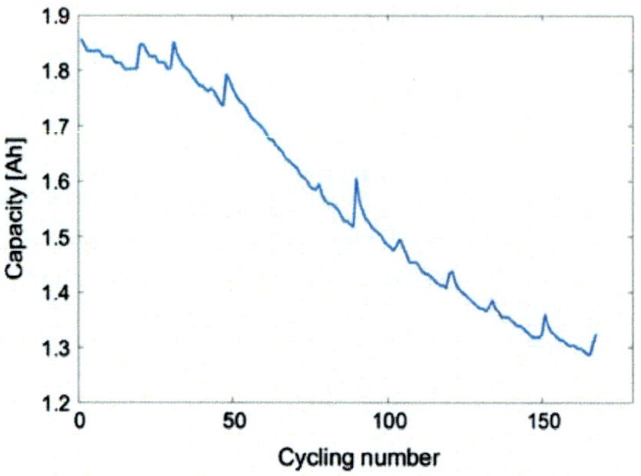

Fig. 2.3 Capacity degradation versus cycling number for battery 05 [2]

2.1.1.7 Coulomb Coefficient

The battery's Coulomb coefficient is calculated by dividing the amount of charges delivered during discharge by the amount of charges held in the battery during the initial charging, which can be expressed as

$$\text{Coulomb coefficient} = \frac{\text{Discharge capacity}}{\text{Charge capacity}} \times 100\%. \tag{2.2}$$

The battery's Coulomb coefficient provides a great deal of information about the level of side reactions that may occur with various battery technologies. An important advantage of Coulomb coefficient as an evaluation parameter is the ability to assess the battery's performance over a short period. The Coulomb coefficient of lithium-ion batteries is similar but never reaches the ideal of 100%. Each cycle loses charge mainly due to side reactions caused by the SEI layer forming or repairing. The loss of capacity at each cycle could also be attributed to electrolyte oxidation, transition metal dissolution, active materials loss from the current collector, and Li plating at the anode. In fact, the Coulomb coefficient of the battery decreases with age.

2.1.2 State Indicators

2.1.2.1 SOC

The SOC of a lithium-ion battery is defined as the ratio of the current available capacity of the battery to its fully charged capacity [3], which is an important parameter to characterize the remaining available power of the battery. It is expressed in percent as shown in Fig. 2.4, from 100% when the battery's available charge is full to 0% when the battery has no available charge. The battery's SOC can be calculated as follows:

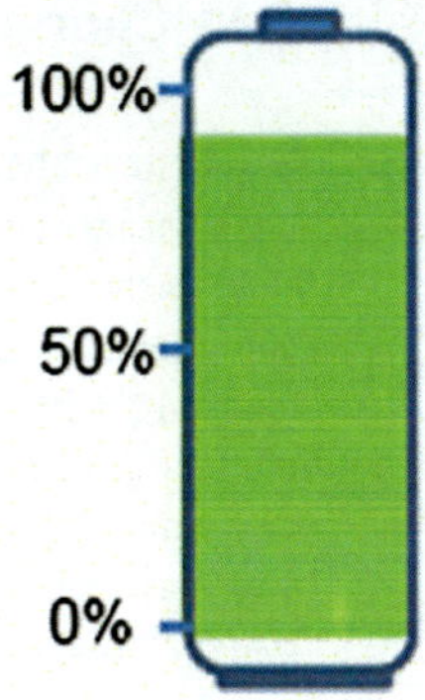

Fig. 2.4 SOC of the battery

$$SOC = SOC_0 - \frac{\eta_0}{C_{rated}} \int I_B \times 100\% \quad (2.3)$$

where SOC_0 denotes the battery's initial SOC, η_0 is the Coulomb coefficient, and C_{rated} is the rated capacity of the battery. As an example, for the cylindrical 18,650 battery with the rated capacity of 2600 mAh, its SOC of 50% means 1300 mAh of energy remaining. Accurately gaining the SOC of the lithium-ion battery can obtain its current remaining power and be utilized in the control algorithm to prevent the lithium-ion battery from overcharging or overdischarging, thus optimizing the battery's performance and extending its lifetime.

2.1.2.2 SOH

The aging of lithium-ion batteries is a long-term gradual process, which is affected by various factors such as temperature, current rate, and cut-off voltage. The state of health (SOH) is an important indicator to characterize the ability of the current battery to store electrical energy relative to the new battery. One of the most commonly used definitions is based on the capacity degradation of the battery as [4]

$$SOH = \frac{C_{aged}}{C_{rated}} \times 100\% \quad (2.4)$$

where C_{aged} and C_{rated} denote the current and rated capacities of the battery, respectively. An SOH of 100% means that the battery capacity has not decayed. During use, the SOH of the battery continues to decrease. Taking the cylindrical 18650 battery with the rated capacity of 2600 mAh as an example, if the SOH of the battery is 80%, it means that the current capacity of the battery is 2080 mAh.

2.2 Charging Objectives and Constraints

The battery charging system is a crucial part of the BMS, yet it is also a major bottleneck for large-scale EV deployments. In the charging control problem, the charging time is the most commonly considered objective, since the reduced charging time can alleviate the mileage anxiety of EV users to a limited extent. To improve battery charging performance such as battery safety and energy conversion efficiency, multiple charging objectives and safety-related constraints are considered. Consequently, designing a reliable battery charging pattern that considers both important objectives and hard constraints plays a key role in the battery management of EVs and other battery-powered systems.

2.2.1 Charging Objectives

As illustrated in Fig. 2.5, the charging time optimization, charging energy loss reduction, temperature rise suppression, and decelerated capacity degradation are important optimization objectives to be considered in charging control for lithium-ion batteries.

2.2.1.1 Charging Time Optimization

For the battery charging management, fast charging speed is one of the most important objectives since long charging time can cause inconvenient use of the battery pack, resulting in consumers' anxiety. This charging control objective aims to minimize the consumed time that the battery is charged from an initial SOC to the desired value SOC_d. Generally, minimizing the charging time of the battery can be expressed as the following equation:

$$\min NT \tag{2.5}$$

where T is the sampling period and N is the sampling step number with $SOC(N) = SOC_d$. Note that the high charging current rates utilized in fast charging strategies will accelerate the degradation of the battery.

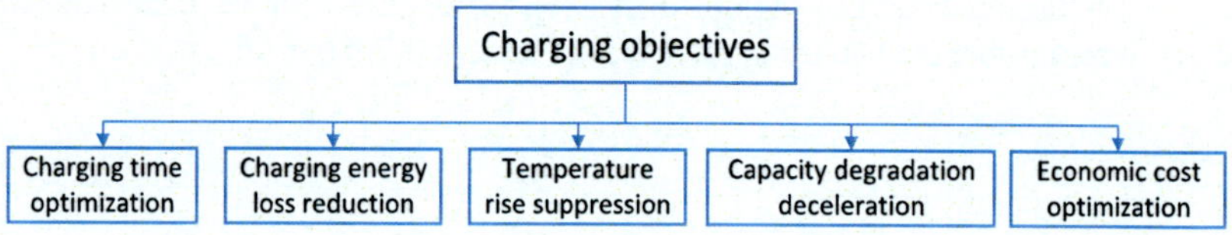

Fig. 2.5 Commonly considered charging objectives

2.2.1.2 Charging Energy Loss Reduction

Energy loss during charging is another crucial indicator commonly considered. Due to the existence of the internal resistance of the battery, some electrical energy is dissipated in the form of heat rather than being converted into chemical energy stored in the battery. It does not only lead to energy waste, but also increases the temperature of the battery. Thereby, it is crucial and necessary to increase the charging efficiency and suppress the temperature rise by reducing the amount of energy loss of the battery during charging.

2.2.1.3 Temperature Rise Suppression

There is heat generated by reactions during battery charging. Since the generated heat cannot be completely dissipated to the environment, it causes the accumulation of heat inside the battery, which results in the high temperature rise, even permanently reducing the battery's capacity and causing thermal runaway of the battery. Hence, it is necessary to take the temperature rise suppression as an important objective in the charging control process. It should be pointed out that the temperature rise of the battery is proportional to the square of its charging current. To ensure that the battery's temperature rise within the normal operating range, the charging current should not be too large.

2.2.1.4 Capacity Degradation Deceleration

Delaying the capacity degradation of the battery is another important goal for charging control, which cannot only maintain battery performance, but also allow the battery to work longer. The capacity loss of the battery is associated with the battery's average SOC, current, and temperature during one cycle. That is why fast charging with a high charging current accelerates the battery capacity decay. It necessaries the charging strategies that can achieve the balance between the charging speed and capacity degradation of the battery.

2.2.1.5 Economic Cost Optimization

Economic cost optimization of battery charging is another necessary and important objective. The electrical price around the world usually fluctuates with the peak or valley of electricity consumption during a day, which is defined as the peak-valley time-of-use (TOU) price. The main economic cost is the money spent on electricity. In practice, it is valuable and necessary to adjust the charging pattern to reduce the economic cost of the consumed electricity, which can save the user's charging electricity bill and improve the user's satisfaction.

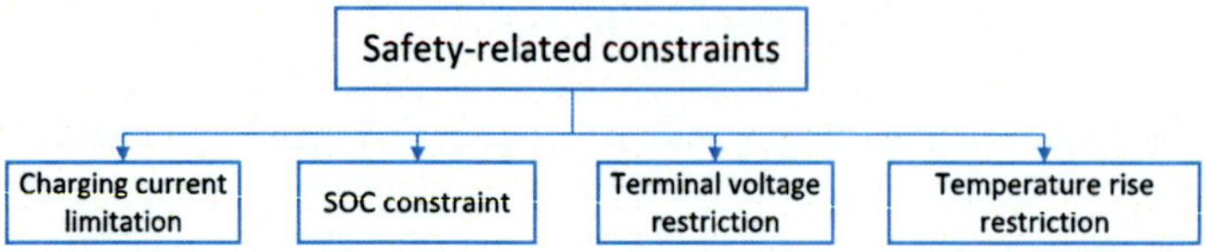

Fig. 2.6 Commonly considered safety-related constraints

2.2.2 Safety-Related Constraints

A number of improper charging behaviours, including overcharging, overcurrent, overvoltage, and overheating, may occur during the charging process. In fact, these behaviors may result in lithium-ion crystallization and precipitation inside the battery, increasing pressure inside the battery, and causing harm to the battery's safety. Hence, the safety-related constraints as shown in Fig. 2.6 should be guaranteed.

2.2.2.1 Charging Current Limitation

The threshold of charging current has a significant role in the battery safety since the excessive current could negatively affect the battery performance or potentially lead to fire throughout the charging process. Accordingly, the charging current of the battery should remain less than its maximum allowable value.

2.2.2.2 SOC Constraint

The SOC is one of the most important states of the battery to indicate its remained energy. In order to prevent the safety hazards caused by overcharging the battery, the battery's SOC must be kept in the range below the upper threshold limit.

2.2.2.3 Terminal Voltage Restriction

Extensive experiments have shown that damage can occur if the charging voltage is elevated above the specification. To prevent damage, the terminal voltage of the battery should be kept not exceed an allowed limit at each sampling time.

2.2.2.4 Temperature Restriction

The temperature is an important indicator that needs to be monitored in the lithium battery management. When the temperature of the battery is out of control, it may cause a fire and even an explosion. Hence, to ensure the safety of the battery, the battery's temperature restriction should be guaranteed. Once the battery's temperature exceeds its limitation, the charging process must be terminated immediately for thermal safety.

References

1. J. Chen, Q. Ouyang, C. Xu, and H. Su, "Neural network-based state of charge observer design for lithium-ion batteries," *IEEE Transactions on Control Systems Technology*, vol. 26, no. 1, pp. 313–320, 2018.
2. Y. Xing, E. W. Ma, K. L. Tsui, and M. Pecht, "An ensemble model for predicting the remaining useful performance of lithium-ion batteries," *Microelectronics Reliability*, vol. 53, no. 6, pp. 811–820, 2013.
3. Y. Hu and S. Yurkovich, "Battery cell state-of-charge estimation using linear parameter varying system techniques," *Journal of Power Sources*, vol. 198, pp. 5043–5049, 2012.
4. C. Lin, A. Tang, and W. Wang, "A review of SOH estimation methods in lithium-ion batteries for electric vehicle applications," *Energy Procedia*, vol. 75, pp. 1920–1925, 2015.

Chapter 3
Lithium-Ion Battery Models

To achieve reasonable charging management for lithium-ion batteries, plenty of model-based strategies have been proposed. For these charging strategies, various types of battery models are used to capture or estimate the battery's in-situ charging states, followed by formulating the specific objective functions to guide charging behaviours. The adopted model can be divided into two main categories, including electrochemical models and equivalent circuit models:

(1) The electrochemical models utilize coupled nonlinear partial differential equations to describe the complex electrochemical reaction mechanism inside the battery during the charging/discharging process.
(2) The equivalent circuit models use circuit elements such as voltage sources, resistors, and capacitors to simulate the dynamic characteristics of the battery without the need to know the complex reaction mechanism inside the battery.

3.1 Electrochemical Models

Electrochemical models can describe the internal reactions of batteries, particularly intercalation and deintercalation of Li^+ in electrode materials, by taking advantage of three transport processes: migration, diffusion, and convection. As shown in Fig. 3.1, the commonly used electrochemical models of the battery include pseudo-two-dimensional model, one-dimensional-based model, and single particle model.

Q. Ouyang and J. Chen, *Advanced Model-Based Charging Control for Lithium-Ion Batteries*, https://doi.org/10.1007/978-981-19-7059-7_3

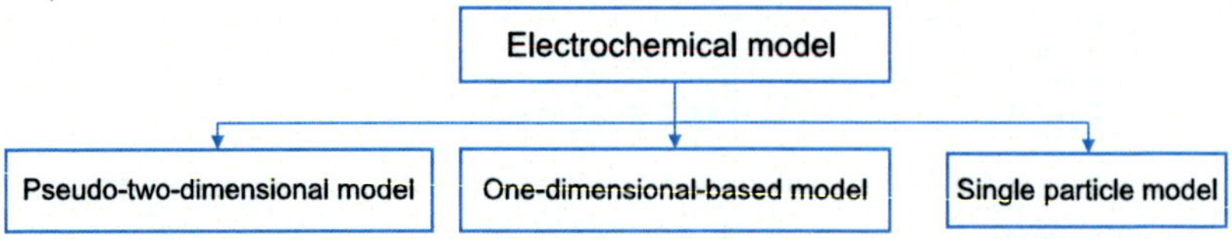

Fig. 3.1 Classification of electrochemical models

Fig. 3.2 Schematic of the P2D lithium-ion battery model [1]

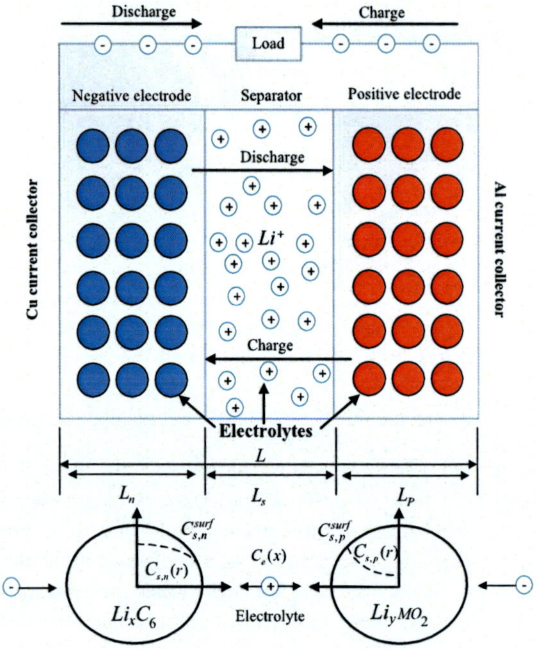

3.1.1 *Pseudo-Two-Dimensional Model*

The pseudo-two-dimensional (P2D) model is one of the most widely used lithium-ion battery models, which is based on a combination of the porous electrode and concentrated solution theories and the kinetics equations [1]. It has been extensively tested and validated that can accurately describe the battery's inner reactions and can make pretty good predictions consistent with empirical data. Figure 3.2 illustrates a schematic of the P2D lithium-ion battery model, where the electrodes are considered as a porous matrix. The governing mathematical equations of the P2D model for both negative and positive electrodes and separators can be calculated as follows:

The Li^+ ions concentration for each electrode follows Fick's law of diffusion for spherical particles as

$$\frac{\partial C_{s,k}(x,r,t)}{\partial t} = \frac{1}{r^2}\frac{\partial}{\partial r}\left(D_{s,k} r^2 \frac{\partial C_{s,k}(x,r,t)}{\partial r}\right) \tag{3.1}$$

where $C_{s,k}(\cdot)$ denotes the solid-state concentration of electrode k ($k = postive, negative$), $D_{s,k}$ is the Li^+ diffusion coefficient in the particle of electrode, and r is the radial coordinate. The liquid-phase Li^+ concentration in the electrolyte is determined by the conservation of Li^+ as:

$$\varepsilon_k \frac{\partial C_{e,k}(x,t)}{\partial t} = \frac{\partial}{\partial x}\left(D_{eff,k}\frac{\partial C_{e,k}(x,t)}{\partial x}\right) + a_k(1 - t_+)J_k(x,t) \tag{3.2}$$

where $C_{e,k}(\cdot)$, $D_{eff,k}$, and $J_k(\cdot)$ denote the electrolyte concentration in region k, the effective diffusion coefficient of Li in electrolyte for region k, and the wall flux of Li^+ on the particle of electrode k, respectively; a_k is the specific surface area of electrode k. The solid-state potential in the electrodes can be calculated based on the Ohm's law as:

$$\sigma_{eff,k}\frac{\partial^2 \phi_{s,k}(x,t)}{\partial x^2} = a_k F J_k(x,t) \tag{3.3}$$

where $\phi_{s,k}(\cdot)$ and $\sigma_{eff,k}$ are the solid-phase potential and effective electronic conductivity of the solid phase of the electrode k, F is the Faraday's constant. The liquid-phase potential in the electrolyte is derived from Kirchhoff's and Ohm's laws as:

$$-\sigma_{eff,k}\frac{\partial \phi_{s,k}(x,t)}{\partial x} - k_{eff,k}\frac{\partial \phi_{e,k}(x,t)}{\partial x} + \frac{2keff,kRT}{F}(1 - t_+)\frac{\partial \ln C_{e,k}}{\partial x} = I \tag{3.4}$$

where $\phi_{e,k}(\cdot)$ and $k_{eff,k}$ are the Electrolyte potential and effective ionic conductivity of the electrolyte in region k, R, T, and t_+ denote the universal gas constant, absolute temperature, and Li^+ transference number in the electrolyte, respectively. The Butler-Volmer kinetics equation describes the pore wall flux of Lithium ions in the electrodes as:

$$\begin{aligned} J_k(x,t) &= K_k(C_{s,k}^{max} - C_{s,k}^{surf})^{\frac{1}{2}}(C_{s,k}^{surf})^{\frac{1}{2}}C_{e,k}^{\frac{1}{2}}[\exp\left(\frac{\frac{1}{2}F\mu_{s,k}(x,t)}{RT}\right) \\ &\quad - \exp\left(-\frac{\frac{1}{2}F\mu_{s,k}(x,t)}{RT}\right)] \\ \mu_{s,k}(x,t) &= \phi_{s,k}(x,t) - \phi_{e,k}(x,t) - U_k \\ V_B(t) &= \phi_{s,k}(0,t) - \phi_{s,k}(L,t) \end{aligned} \tag{3.5}$$

where V_B and U_k are the battery's terminal voltage and open circuit potential of electrode k, $\mu_{s,k}(\cdot)$ is overpotential of electrode k, $C_{s,k}^{max}$ and $C_{s,k}^{surf}$ denote the maximum concentration of Li^+ in and concentration of on the surface of the particle of electrode k, respectively. The liquid-phase Lithium ions concentration in the separator is determined by the conservation of Lithium ions as:

$$\varepsilon_k \frac{\partial C_{e,k}(x,t)}{\partial t} = \frac{\partial}{\partial x}\left(D_{eff,k}\frac{\partial C_{e,k}(x,t)}{\partial x}\right). \tag{3.6}$$

The liquid-phase potential (ϕ_e) in the separator is derived from Kirchhoff's and Ohm's laws as:

$$-k_{eff,k}\frac{\partial\phi_{e,k}(x,t)}{\partial x} + \frac{2k_{eff,k}RT}{F}(1-t_+)\frac{\partial \ln C_{e,k}}{\partial x} = I. \tag{3.7}$$

It should be pointed out that it is hard to obtain a comprehensive analytical solution to the governing equations for the P2D model. Consequently, various numerical methods, including the finite-difference method and the orthogonal projections [2], have been utilized to estimate the model parameters.

3.1.2 One-Dimensional Model

The one-dimensional electrochemical model of the battery normally involves four key parts, namely: the porous negative electrode, the separator, the porous positive electrode, and the electrolyte [3]. Accordingly, the discharge process occurs when lithium ions diffuse through the negative electrode's solid particles to the surface. Upon reaching the surface, the ions are transformed into the electrolyte by an electrochemical reaction. Following their entry into the electrolyte solution, the ions travel to the positive electrode, where they react and are intercalated into the solid particles once more. In the charging process, the liquid phase moves from the positive electrode to the negative electrode by flowing lithium ions from the solid phase. Since the separator acts as an electrically insulating barrier between the electrodes, the electrons involved in these charging and discharging processes flow between the two compartments through an external circuit [4]. Subsequently, the electrochemical processes can be described by a model containing six state variables including electric potential $P_x(x,t)$ in the solid electrode, the electrolyte's electrical potential $P_e(x,t)$, the active material's lithium ion concentration concerning positive and negative electrodes $C_s(x,r,t)$, the lithium ion concentration in the electrolyte $C_e(x,t)$, the electrolyte's ionic current $i_e(x,t)$, as well as the molar ion fluxes among the active material within the electrodes and electrolyte $j_n(x,t)$. The governing mathematical equations can be calculated as [3].

The solid electrodes' potential:

$$\frac{\partial P_s(x,t)}{\partial x} = \frac{i_e(x,t)-I(t)}{\sigma}. \tag{3.8}$$

The electrolyte's potential:

$$\frac{\partial P_s(x,t)}{\partial x} = -\frac{i_e(x,t)}{\kappa} + \frac{2RT}{F}(1-t_c^0)\left(1+\frac{\partial \ln f_{c/a}}{\partial \ln C_e}(x,t)\right)\frac{\partial C_e(x,t)}{\partial x}. \tag{3.9}$$

The solid particles' lithium ion concentration:

$$\frac{\partial C_s(x,r,t}{\partial t} = \frac{1}{r^2}\frac{\partial}{\partial r}\left(D_s r^2\frac{\partial C_s(x,r,t)}{\partial r}\right). \tag{3.10}$$

The electrolyte's lithium ion concentration:

$$\frac{\partial C_e(x,t}{\partial x} = \frac{\partial}{\partial x}\left(D_e \frac{\partial C_e(x,t)}{\partial x}\right) + \frac{1}{F\varepsilon_e}\frac{\partial t_a^0 i_e(x,t)}{\partial x}. \tag{3.11}$$

The electrolyte's current:

$$\frac{\partial i_e(x,t)}{\partial x} = aFj_n(x,t). \tag{3.12}$$

The molar flux driven by the Butler-Volmer equation:

$$j_n(x,t) = \frac{i_0(x,t)}{F}\left[\exp\left(\frac{\alpha_a F}{RT}\eta_s(x,t)\right) - \exp\left(\frac{-\alpha_c F}{RT}\eta_s(x,t)\right)\right] \tag{3.13}$$

where $\eta_s(\cdot)$ and $i_0(\cdot)$ stand for the intercalation reaction's over-potential and exchange current density, respectively, and can be formulated as:

$$\begin{aligned} \eta_s(x,t) &= \theta_s(x,t) - \theta_e(x,t) - U(C_{ss}(x,t)) - FR_f j_n(x,t) \\ i_0(x,t) &= r_{eff} C_e(x,t)^{\alpha_a}(C_{s_{max}} - C_{ss}(x,t))^{\alpha_a} C_{ss}(x,t)^{\alpha_c}. \end{aligned} \tag{3.14}$$

3.1.3 Single Particle Model

The single particle model is a simplification of the P2D model with less computational complexity, which ignores the electrolyte characteristics [5]. The single particle model is based on two primary assumptions. First, each electrode is modelled according to two spherical particles where intercalation and deintercalation phenomena take place. Second, the variation of electrolyte concentration and different potentials are ignored. The schematic of the single particle model is depicted as in Fig. 3.3. Its governing equations include the solid-state concentration (3.1) as well as the Butler-Volmer kinetics (3.5) for positive and negative electrodes [6].

The single particle model has the advantage of reasonable computation complexity, which is suitable for a variety of purposes, including online state estimation and charging control of the battery. However, it also has its drawbacks, such as it ignores the electrolyte characteristics, while the thick electrodes and high discharge rates necessitate customized settings based on these characteristics [7].

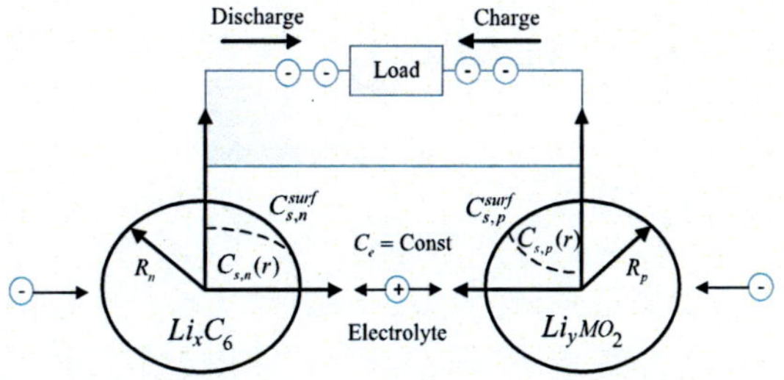

Fig. 3.3 Schematic of the single particle model [1]

3.2 Equivalent Circuit Models

The equivalent circuit model can be thought of as a performance model that creates an electric circuit replicating the lithium-ion battery's voltage-current characteristics, SOC dynamics, etc., by using a series (one or more) of parallel combinations of the voltage source, resistance, capacitance, and other electronic components. Due to obvious merits such as simple structure and being flexible, they can be easily implemented in real applications, making the equivalent circuit models become the mainstream candidates in the existing designs of model-based charging strategies for lithium-ion batteries. As shown in Fig. 3.4, the commonly used equivalent circuit models include Rint model, Thevenin model, and Partnership for New Generation of Vehicles (PNGV) model, etc. [8].

3.2.1 Rint Model

The Rint model is illustrated in Fig. 3.5, which is composed of a power source and a resistance [9]. In Fig. 3.5, V_{OC} and R_{int} denote the OCV and internal resistance of the battery, V_B and I_B are the battery's terminal voltage and current, respectively. Its mathematical model can be deduced as follows:

$$\begin{aligned} \dot{SOC} &= -\frac{1}{C_b} I_B \\ V_B &= V_{oc}(SOC) - R_{int} I_B \end{aligned} \tag{3.15}$$

where SOC and C_b stand for the SOC and capacity of the battery. It should be noted that the Rint model is one of the simplest equivalent models, which has a relatively low accuracy.

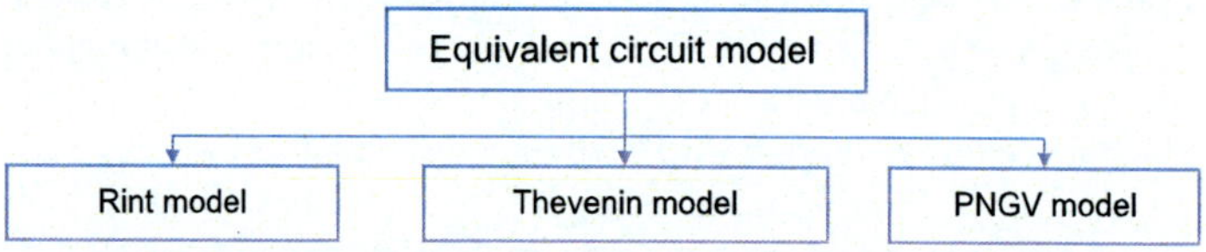

Fig. 3.4 Empirical-model-based charging classification

Fig. 3.5 A typical Rint model

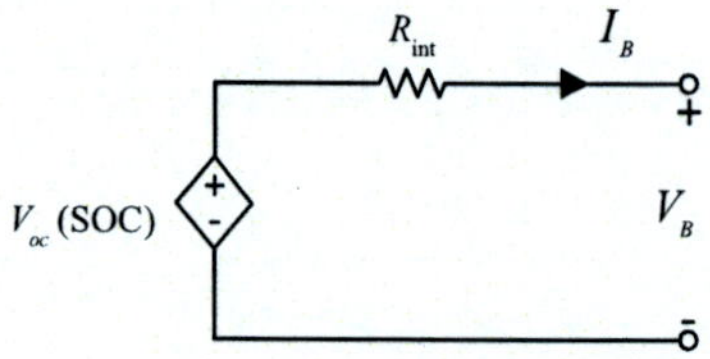

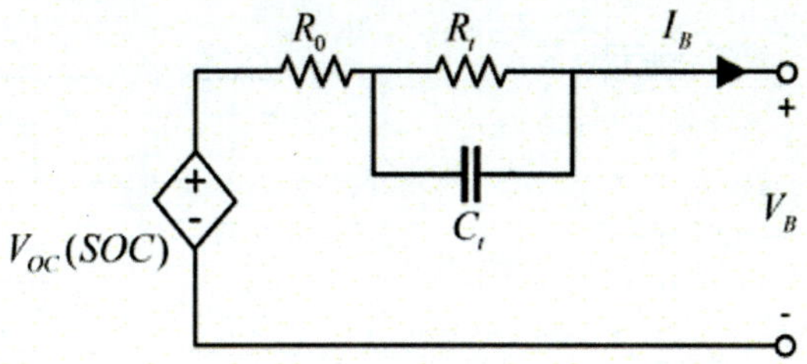

Fig. 3.6 Thevenin model with 1 RC branch

3.2.2 *Thevenin Model*

As the improvement of the Rint model, the Thevenin model takes the polarization effect of the battery into consideration, where the polarization effect causes the electrode potential to deviate from the equilibrium point during charging and discharging and can generally be divided into electrochemical polarization and concentration polarization. In the Thevenin model, as illustrated in Fig. 3.6, a RC network (R_t, C_t) is utilized to simulate the polarization reaction inside the battery to improve the model accuracy [10]. Its mathematical model can be deduced as:

$$\begin{aligned} S\dot{O}C &= -\tfrac{1}{C_b} I_B \\ \dot{V}_t &= -\tfrac{1}{R_t C_t} V_t + \tfrac{1}{C_t} I_B \\ V_B &= V_{OC}(SOC) - R_0 I_B - V_t \end{aligned} \tag{3.16}$$

where V_t denotes the voltage across the capacitor C_t.

Usually, more RC networks can be utilized in the Thevenin model that can better describe the dynamic characteristics of the battery. In fact, with the increase of the RC network, the accuracy of the model can be correspondingly improved, but it leads to an increase in the calculation complexity of the model. In practice, the 2 RC equivalent circuit model is the commonly used [11] as shown in Fig. 3.7. Its model can be expressed as:

$$\begin{aligned} S\dot{O}C &= -\tfrac{1}{C_b} I_B \\ \dot{V}_s &= -\tfrac{1}{R_s C_s} V_s + \tfrac{1}{C_s} I_B \\ \dot{V}_f &= -\tfrac{1}{R_f C_f} V_f + \tfrac{1}{C_f} I_B \\ V_B &= V_{OC}(SOC) - R_0 I_B - V_s - V_f \end{aligned} \tag{3.17}$$

where V_s and V_f denote the voltage across the capacitor C_s and C_f, respectively.

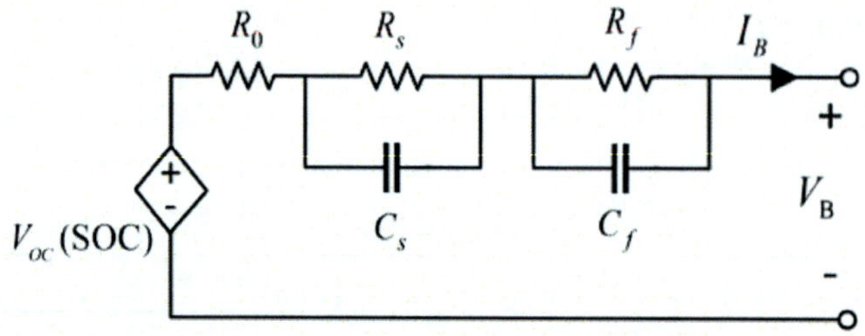

Fig. 3.7 Thevenin model with 2 RC branch

Fig. 3.8 PNGV model

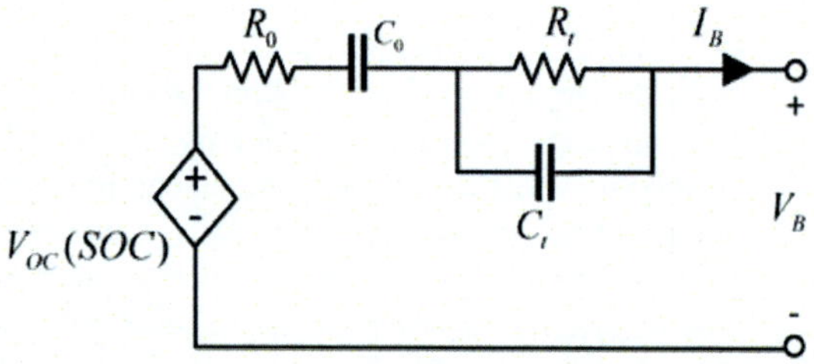

3.2.3 *PNGV Model*

The PNGV model [12] adds the influence of the load change on the battery's OCV during use based on the Thevenin model. The model structure is shown in Fig. 3.8, which can be considered as a capacitor C_0 is added in the Thevenin model (as in Fig. 3.5). Its mathematical model can be calculated as follows:

$$\begin{aligned}
&S\dot{O}C = -\frac{1}{C_b} I_B \\
&\dot{V}_0 = \frac{1}{C_0} I_B \\
&\dot{V}_t = -\frac{1}{R_t C_t} V_t + \frac{1}{C_t} I_B \\
&V_B = V_{OC}(SOC) - V_0 - R_0 I_B - V_t
\end{aligned} \tag{3.18}$$

where V_0 denotes the voltage across the capacitor C_0. Compared with the Thevenin model, the PNGV model has higher accuracy, but the corresponding computational complexity is also increased.

References

1. Ali Jokar, Barzin Rajabloo, Martin Dsilets, Marcel Lacroix, "Review of simplified pseudo-two-dimensional models of lithium-ion batteries," *Journal of Power Sources*, vol. 327, pp. 44–55, 2016.

2. M.D. Beeney, *Lithium Ion Battery Modeling Using Orthogonal*, Ph.D. dissertation, The Pennsylvania State University, 2013.
3. N. A. Chaturvedi, R. Klein, J. Christensen, J. Ahmed, and A. Kojic, "Algorithms for advanced battery-management systems," *IEEE Control Systems Magazine*, vol. 30, no. 3, pp. 49–68, 2010.
4. K. Smith, C. Y. Wang, "Solid-state diffusion limitations on pulse operation of a lithium ion cell for hybrid electric vehicles," *Journal of Power Sources*, vol. 161, no. 1, pp. 628–639, 2006.
5. H. Perez, S. Dey, X. Hu, and S. Moura, "Optimal charging of li-ion batteries via a single particle model with electrolyte and thermaldynamics," *Journal ofThe Electrochemical Society*, vol. 164, pp. A1679, 2017.
6. S. Santhanagopalan, Q. Guo, P. Ramadass, and R. E. White, "Review of models for predicting the cycling performance of lithium ion batteries," *Journal of Power Sources*, vol. 156, no. 2, pp. 620–628, 2006.
7. V. Ramadesigan, P. W. Northrop, S. De, S. Santhanagopalan, R. D. Braatz, and V. R. Subramanian, "Modeling and simulation of lithium-ion batteries from a systems engineering perspective," *Journal of The Electrochemical Society*, vol. 159, no. 3, pp. R31, 2012.
8. X. Zhang, W. Zhang, and G. Lei, "A review of li-ion battery equivalent circuit models," *Transactions on Electrical and Electronic Materials*, vol. 17, no. 6, pp. 311–316, 2016.
9. D. Andrea, *Battery Management Systems for Large Lithium-Ion Battery Packs*, London: Artech House, 2010.
10. M. A. Roscher, O. S. Bohlen, and D. U. Sauer, "Reliable state estimation of multicell lithium-ion battery systems," *IEEE Transactions on Energy Conversion*, vol. 26, no. 3, pp. 737–743, 2011.
11. H. He, R. Xiong, X. Zhang, F. Sun, and J. Fan, "State-of-charge estimation of the lithium-ion battery using an adaptive extended Kalman filter based on an improved Thevenin model," *IEEE Transactions on Vehicular Technology*, vol. 60, no. 4, pp. 1461–1469, 2011.
12. X. Liu, W. Li, and A. Zhou, "PNGV equivalent circuit model and SOC estimation algorithm for lithium battery pack adopted in AGV vehicle," *IEEE Access*, vol. 6, pp. 23639–23647, 2018.

Chapter 4
Neural Network-Based State of Charge Observer for Lithium-Ion Batteries

For lithium-ion batteries, the SOC is one of the most crucial parameters of the BMS, which is usually used to optimize the operation and extend the life of the battery. When there exists a large SOC estimation error, it can easily lead to overdischarging and overcharging of the battery, which can cause irreversible damage to the battery and even lead to an explosion in severe cases. Therefore, an efficient and high-precision SOC estimation method is needed for the lithium-ion battery.

The currently available SOC estimation methods fall into two main types, non-model-based and model-based [1]. The non-model-based methods do not rely on mathematical models that capture the battery dynamics [2]. One of the mainstream methods is ampere-hour counting [3], where the SOC is estimated by accumulating and calculating the energy (in Ampere-hours) of charging or discharging the battery. Although this method is easy to be implemented, the calculation error may accumulate over time, which eventually leads to a large deviation of the estimated SOC from the actual value. Therefore, in order to suppress this error, it is necessary to perform frequent calibrations. Another widely used SOC estimation method is called voltage translation [4], which infers the relationship between SOC and OCV by querying a predetermined table. Although the error in the estimated value is reduced, this method must disconnect the battery from the external circuit and let the battery rest for a long time to measure its OCV accurately.

Since the SOC estimation accuracy can be improved by using the battery's dynamics described by the battery model, model-based estimation methods have received great attention in recent years. In this chapter, based on an improved Thevenin model with 2 RC branches, a radial basis function (RBF) neural network-based observer is proposed for the SOC estimation of the battery [5].

4.1 Battery Model

An improved Thevenin model [6] is utilized to describe the dynamics of the battery. As illustrated in Fig. 4.1, it includes two interrelated sub-circuits that interact

Q. Ouyang and J. Chen, *Advanced Model-Based Charging Control for Lithium-Ion Batteries*, https://doi.org/10.1007/978-981-19-7059-7_4

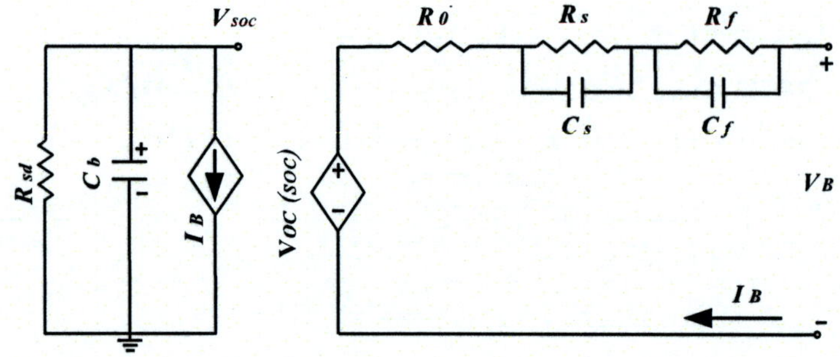

Fig. 4.1 Battery equivalent circuit model

with each other through a voltage-controlled voltage source and a current-controlled current source.

- The circuit on the left in Fig. 4.1 is used to simulate the battery's characteristics of SOC and remaining runtime. C_b denotes the full charge stored inside the battery, and R_{sd} indicates the self-discharge resistance. C_b and R_{sd} are used to show the self-discharge characteristics of the battery. V_{SOC} quantitatively represents the SOC of the battery, and V_{SOC} of 0–1 V corresponds to 0–100% of the SOC. In this model, the voltage-controlled voltage source provides a nonlinear mapping from the SOC of the battery to its OCV. It can be expressed as

$$V_{OC} = f(V_{SOC}) \tag{4.1}$$

where $f(\cdot)$ is a nonlinear function. It is assumed that its first order derivative exists.
- The circuit on the right in Fig. 4.1 is utilized to simulate the battery's transient response and V-I curve. The resistor R_0 is used to describe the battery's charge/discharge energy loss. The RC networks (R_s, C_s) and (R_f, C_f) are used to describe the battery's short-term and long-term transient responses. V_B represents the terminal voltage of the battery. I_B denotes the battery's current, which is positive/negative when the battery is in discharge/charge mode, respectively.

All parameters in this model are considered as the functions of the battery's SOC, so that this model has a higher accuracy compared to the model with constant parameters. In practical applications, to simplify the battery model, C_b can be treated as the rated capacity of the battery, and R_{sd} can be simplified by a constant large resistance when the temperature of the battery varies in a small range. The dynamics of the voltages across the capacitors C_b, C_s and C_f are denoted as V_{SOC}, V_s and V_f, which can be expressed as follows

$$\begin{aligned}
\dot{V}_{SOC} &= -\tfrac{1}{R_{sd}C_b}V_{SOC} - \tfrac{1}{C_b}I_B \\
\dot{V}_s &= -\tfrac{1}{R_s(V_{SOC})C_s(V_{SOC})}V_s + \tfrac{1}{C_s(V_{SOC})}I_B \\
\dot{V}_f &= -\tfrac{1}{R_f(V_{SOC})C_f(V_{SOC})}V_f + \tfrac{1}{C_f(V_{SOC})}I_B.
\end{aligned} \tag{4.2}$$

The terminal voltage V_B can be represented as

$$V_B = V_{OC} - R_0(V_{SOC})I_B - V_f - V_s. \tag{4.3}$$

As the derivative of the battery's current I_B can be ignored compared with other time-constants in a sampling period, based on (4.1)–(4.3), the derivative of the terminal voltage can be determined as follows:

$$\begin{aligned}\dot{V}_B &= \frac{\partial V_{OC}}{\partial V_{SOC}}\dot{V}_{SOC} - \dot{V}_f - \dot{V}_s \\ &= \frac{V_f}{R_f(V_{SOC})C_f(V_{SOC})} - \frac{V_f}{R_s(V_{SOC})C_s(V_{SOC})} - \frac{V_B}{R_s(V_{SOC})C_s(V_{SOC})} \\ &\quad - [\frac{R_0(V_{SOC})}{C_s(V_{SOC})R_s(V_{SOC})} + \frac{1}{C_f(V_{SOC})} + \frac{1}{C_s(V_{SOC})} + \frac{\dot{f}(V_{SOC})}{C_b}]I_B \\ &\quad + \frac{f(V_{SOC})}{R_s(V_{SOC})C_s(V_{SOC})} - \frac{\dot{f}(V_{SOC})}{R_{sd}C_b}V_{SOC}\end{aligned} \tag{4.4}$$

where $\dot{f}(V_{SOC})$ denotes the derivative of $f(V_{SOC})$ with respect to V_{SOC}. With the model bias and unknown disturbance considered, based on (4.2) and (4.4), the battery model can be rewritten as follows:

$$\begin{aligned}\dot{x} &= A(x)x + g(x,u) + \varphi(x,u) \\ y &= Cx\end{aligned} \tag{4.5}$$

where $x(t) \triangleq [V_{SOC}, V_f, V_s, V_B]^T \in \mathbb{R}^4$ is the state vector of the battery model. $y(t) \triangleq V_B \in \mathbb{R}$ and $u(t) \triangleq I_B \in \mathbb{R}$ are the output and input of the battery model, respectively. $\varphi(x,u) \in \mathbb{R}^4$ denotes the part of the model errors and unmeasured disturbances. $A(x) \in \mathbb{R}^{4\times4}$, $g(x,u) \in \mathbb{R}^4$ and $C \in \mathbb{R}^{1\times4}$ are defined as follow

$$A(x) = \begin{bmatrix} -a & 0 & 0 & 0 \\ 0 & -h_1(x_1) & 0 & 0 \\ 0 & 0 & -h_2(x_1) & 0 \\ 0 & h_3(x_1) & 0 & -h_2(x_1) \end{bmatrix} \tag{4.6}$$

$$g(x,u) = \begin{bmatrix} -bu \\ g_1(x_1)u \\ g_2(x_1)u \\ g_3(x_1)u + g_4(x_1) \end{bmatrix} \tag{4.7}$$

$$\varphi(x,u) = [0\ 0\ 0\ \psi(x,u)]^T \tag{4.8}$$

$$C = [0\ 0\ 0\ 1] \tag{4.9}$$

with

$$a = \frac{1}{R_{sd}C_b},\quad b = \frac{1}{C_b},\quad h_1(x_1) = \frac{1}{R_f(V_{SOC})C_f(V_{SOC})}$$

$$h_2(x_1) = \frac{1}{R_s(V_{SOC})C_s(V_{SOC})},\quad h_3(x_1) = \frac{1}{R_f(V_{SOC})C_f(V_{SOC})} - \frac{1}{R_s(V_{SOC}))C_s(V_{SOC})}$$

$$g_1(x_1) = \frac{1}{C_f(V_{SOC})},\quad g_2(x_1) = \frac{1}{C_s(V_{SOC})}$$

$$g_3(x_1) = -\frac{R_0(V_{SOC})}{C_s(V_{SOC})R_s(V_{SOC})} - \frac{1}{C_f(V_{SOC})} - \frac{1}{C_s(V_{SOC})} - \frac{\dot{f}(V_{SOC})}{C_b}$$

$$g_4(x_1) = \frac{f(V_{SOC})}{R_s(V_{SOC})C_s(V_{SOC})} - \frac{\dot{f}(V_{SOC})}{R_{sd}C_b}V_{SOC}.$$

The formulation in (4.5) illustrates a single-input single-output nonlinear system, which is complicated and hard to be analyzed because of its high nonlinearity. Note that the resistors and capacitors in the equivalent circuit model are expected to be positive and bounded. The battery charging/discharging currents are bounded due to the protection of the BMS. All states of the battery system are bounded with the non-extreme current of the battery. It is then obtained that $h_1(\cdot)$ and $h_2(\cdot)$ are both positive and bounded in (4.6) and $u(t)$, $x(t)$, $h_3(\cdot)$, $g_i(\cdot)$ $(1 \leq i \leq 4)$ are bounded.

4.2 Neural Network-Based Nonlinear Observer Design for SOC Estimation

4.2.1 Neural Network-Based Nonlinear Observer Design

In the nonlinear battery model (4.5), the state $x_1(t)$ represents the SOC of the battery, but is not directly measurable. The only measurable state is the terminal voltage of the battery, which is the output $y(t)$ of the model. To estimate the SOC of the battery accurately, a nonlinear observer using a RBF neural network for online approximation of uncertainties in the battery model is proposed as follows:

$$\dot{\hat{x}} = A(\hat{x})\hat{x} + g(\hat{x}, u) + \hat{\varphi}(\hat{x}, u) + L(y - C\hat{x}) \tag{4.10}$$

where $\hat{x}(t) \in \mathbb{R}^4$ is an estimate of $x(t)$, $L \triangleq [l_1, l_2, l_3, l_4]^T$ is the observer gain vector to be designed, and $\hat{\varphi}(\cdot)$ is the estimation of $\varphi(\cdot)$ in (4.5) by utilizing RBF neural network. $\hat{\psi}(\cdot)$ is a RBF neural network to estimate the modeling error and unmeasurable disturbances as

$$\hat{\psi} = W^T S(\hat{z}) \tag{4.11}$$

where the input vector $\hat{z} = [\hat{x}^T, u]^T$, $W = [w_1, w_2, \ldots, w_n]^T$ is the weight vector, n denotes the number of nodes in the neural network, and $S(\hat{z}) = [s_1(\hat{z}), \ldots, s_n(\hat{z})]^T$ is the activation function vector with $s_i(\hat{z})(1 \leq i \leq n)$ satisfying the following conditions

$$s_i(\hat{z}) = \exp[\frac{-(\hat{z}-\mu_i)^T(\hat{z}-\mu_i)}{\eta_i^2}] \tag{4.12}$$

where $\mu_i \in \mathbb{R}^5$ is the center of the receptive field and $\eta_i \in \mathbb{R}$ is the width of the Gaussian function. The adaptive law of the weight vector can be adopted as follows:

$$\dot{W} = \Gamma[S(\hat{z})(y - C\hat{x}) - K_w W] \tag{4.13}$$

where $\Gamma = \Gamma^T \in \mathbb{R}^{n\times n}$ and $K_w \in \mathbb{R}^{n\times n}$ are positive definite constant matrices to be designed. It has been proven that an RBF neural network with a sufficient number of neural nodes can approximate any continuous function [7]. Based on this approximation property of the RBF neural network, the uncertainty $\psi(\cdot)$ can then be substituted with

$$\psi(x, u) = W^{*T} S(z) + \xi \tag{4.14}$$

where W^* is the ideal constant weight vector, the input vector is $z = [x^T, u]^T$, ξ is the approximation error, and it can be assumed that $|\xi| \le \xi_N$. Since $\psi(\cdot)$ is bounded, the ideal constant weights W^* are also bounded that satisfies

$$\|W^*\| \le W_M \tag{4.15}$$

where W_M is a positive constant. Then, the RBF neural network-based nonlinear observer in (4.10) can be rewritten as follows:

$$\begin{aligned}
\dot{\hat{x}}_1 &= -a\hat{x}_1 - bu + l_1(x_4 - \hat{x}_4) \\
\dot{\hat{x}}_2 &= -h_1(\hat{x}_1)\hat{x}_2 + g_1(\hat{x}_1)u + l_2(x_4 - \hat{x}_4) \\
\dot{\hat{x}}_3 &= -h_2(\hat{x}_1)\hat{x}_3 + g_2(\hat{x}_1)u + l_3(x_4 - \hat{x}_4) \\
\dot{\hat{x}}_4 &= h_3(\hat{x}_1)\hat{x}_2 - h_2(\hat{x}_1)\hat{x}_4 + g_3(\hat{x}_1)u + g_4(\hat{x}_1) \\
&\quad + W^T S(\hat{x}, u) + l_4(x_4 - \hat{x}_4) \\
\dot{W} &= \Gamma[S(\hat{x}, u)(y - C\hat{x}) - K_w W]
\end{aligned} \tag{4.16}$$

where $h_i(\cdot)(1 \le i \le 3)$ and $g_i(\cdot)(1 \le i \le 4)$ are all bounded. Since $h_1(\cdot)$ and $h_2(\cdot)$ are positive, we can define $-|h_1(\cdot)| \le -d_1$, $-|h_2(\cdot)| \le -d_2$ and $|h_3(\cdot)| \le d_3$ for all positive constants $d_i(1 \le i \le 3)$. Based on (4.5), (4.14) and (4.16), the estimation error $\tilde{x}(t) \triangleq x(t) - \hat{x}(t)$ can be deduced as

$$\begin{aligned}
\dot{\tilde{x}}_1 &= -a\tilde{x}_1 - l_1\tilde{x}_4 \\
\dot{\tilde{x}}_2 &= -h_1(x_1)x_2 + h_1(\hat{x}_1)\hat{x}_2 + (g_1(x_1) - g_1(\hat{x}_1))u - l_2\tilde{x}_4 \\
\dot{\tilde{x}}_3 &= -h_2(x_1)x_3 + h_2(\hat{x}_1)\hat{x}_3 + (g_2(x_1) - g_2(\hat{x}_1))u - l_3\tilde{x}_4 \\
\dot{\tilde{x}}_4 &= h_3(x_1)x_2 - h_3(\hat{x}_1)\hat{x}_2 - h_2(x_1)x_4 + h_2(\hat{x}_1)\hat{x}_4 \\
&\quad + (g_3(x_1) - g_3(\hat{x}_1))u + (g_4(x_1) - g_4(\hat{x}_1)) \\
&\quad + W^{*T} S(x, u) + \xi - W^T S(\hat{x}, u) - l_4\tilde{x}_4.
\end{aligned} \tag{4.17}$$

Since $h_i(\cdot)(1 \le i \le 3), g_i(\cdot)(1 \le i \le 4), x_i(t)(1 \le i \le 5)$ and $u(t)$ are all bounded, It can be inferred that

$$\begin{aligned}
&|(h_1(\hat{x}_1) - h_1(x_1))x_2 + (g_1(x_1) - g_1(\hat{x}_1))u| \le M_1 \\
&|(h_2(\hat{x}_1) - h_2(x_2))x_3 + (g_2(x_1) - g_2(\hat{x}_1))u| \le M_2 \\
&|(h_3(x_1) - h_3(\hat{x}_1))x_2 + (h_2(\hat{x}_1) - h_2(x_1))x_4| \\
&\quad + |(g_3(x_1) - g_3(\hat{x}_1))u + g_4(x_1) - g_4(\hat{x}_1)| \le M_3
\end{aligned} \tag{4.18}$$

where $M_i(1 \le i \le 3)$ are positive bounded constants. By simultaneously subtracting and adding $W^{*T}S(\hat{x}, u)$, it can be obtained that

$$W^{*T}S(x, u) + \xi - W^T S(\hat{x}, u) = -\tilde{W}^T S(\hat{x}, u) + \varepsilon \tag{4.19}$$

where

$$\begin{aligned}
\tilde{W} &= W - W^* \\
\varepsilon &= W^{*T}(S(x, u) - S(\hat{x}, u)) + \xi.
\end{aligned} \tag{4.20}$$

Note that ε in (4.20) is a bounded term that satisfies $|\varepsilon| \le \varepsilon_M (\varepsilon_M > 0)$.

4.2.2 Convergence Analysis

To prove the convergence of the designed RBF neural network-based nonlinear observer (4.10), a Lyapunov function $V(t) \in \mathbb{R}$ is selected as follows:

$$V = \tfrac{1}{2}p_1\tilde{x}_1^2 + \tfrac{1}{2}\tilde{x}_2^2 + \tfrac{1}{2}\tilde{x}_3^2 + \tfrac{1}{2}\tilde{x}_4^2 + \tfrac{1}{2}\tilde{W}^T\Gamma^{-1}\tilde{W} \tag{4.21}$$

where p_1 is a positive constant and Γ^{-1} is the inverse matrix of Γ in (4.13), which is positive definite.

Theorem 4.1 *If the observer gain l_4 satisfies*

$$l_4 > \tfrac{2p_1l_1^2}{a} - \tfrac{3(d_3+|l_2|)^2}{d_1} - \tfrac{3l_3^2}{d_2}, \tag{4.22}$$

the SOC estimation error of the proposed RBF neural network-based nonlinear observer is uniformly bounded.

Proof Since W^* is a constant weight vector, it can be inferred that

$$\dot{\tilde{W}} = \dot{\hat{W}}. \tag{4.23}$$

From (4.18)–(4.20), the derivative of $V(t)$ in (4.21) is

$$\begin{aligned}\dot{V} &= p_1\tilde{x}_1\dot{\tilde{x}}_1 + \tilde{x}_2\dot{\tilde{x}}_2 + \tilde{x}_3\dot{\tilde{x}}_3 + \tilde{x}_4\dot{\tilde{x}}_4 + \tilde{W}^T\Gamma^{-1}\dot{\hat{W}} \\ &\leq -ap_1\tilde{x}_1^2 + |l_1 p_1\tilde{x}_1\tilde{x}_4| - d_1\tilde{x}_2^2 + M_1|\tilde{x}_2| + |d_3\tilde{x}_2\tilde{x}_4| \\ &+ |l_2\tilde{x}_2\tilde{x}_4| - d_2\tilde{x}_3^2 + M_2|\tilde{x}_3| + |l_3\tilde{x}_3\tilde{x}_4| - d_2\tilde{x}_4^2 \\ &+ M_3|\tilde{x}_4| - l_4\tilde{x}_4^2 + \varepsilon_M|\tilde{x}_4| - \tilde{W}^T K_w W.\end{aligned} \tag{4.24}$$

According to Lemma A.17 in [8], the following inequality can be established

$$\begin{aligned}&-ap_1\tilde{x}_1^2 - |l_1 p_1\tilde{x}_1\tilde{x}_4| \leq -\tfrac{ap_1}{2}\tilde{x}_1^2 + \tfrac{2p_1l_1^2\tilde{x}_4^2}{a} \\ &-d_1\tilde{x}_2^2 + M_1|\tilde{x}_2| + |d_3\tilde{x}_2\tilde{x}_4| + |l_2\tilde{x}_2\tilde{x}_4| \leq -\tfrac{d_1}{3}\tilde{x}_2^2 + \tfrac{3M_1^2}{d_1} + \tfrac{3(d_3+|l_2|)^2}{d_1}\tilde{x}_4^2 \\ &-d_2\tilde{x}_3^2 + M_2|\tilde{x}_3| + |l_3\tilde{x}_3\tilde{x}_4| \leq -\tfrac{d_2}{3}\tilde{x}_3^2 + \tfrac{3M_2^2}{d_2} + \tfrac{3l_3^2}{d_2}\tilde{x}_4^2 \\ &-d_2\tilde{x}_4^2 + M_3|\tilde{x}_4| - l_4\tilde{x}_4^2 + \varepsilon_M|\tilde{x}_4| \leq -l_4\tilde{x}_4^2 + \tfrac{(M_3+\varepsilon_M)^2}{d_2}.\end{aligned} \tag{4.25}$$

By completion of squares, it can be deduced that

$$\begin{aligned}-\tilde{W}^T K_w W &= -\tilde{W}^T K_w(\tilde{W} + W^*) \\ &\leq -\frac{\|K_w\|\,\|\tilde{W}\|^2}{2} + \frac{\|K_w\|\,W_M^2}{2}.\end{aligned} \tag{4.26}$$

Based on (4.25) and (4.26), (4.24) is deduced as:

$$\dot{V} \leq -\tfrac{ap_1}{2}\tilde{x}_1^2 - \tfrac{d_1}{3}\tilde{x}_2^2 - \tfrac{d_2}{3}\tilde{x}_3^2 - \alpha\tilde{x}_4^2 - \tfrac{\|K_w\|\|\tilde{W}\|^2}{2} + M \tag{4.27}$$

with

$$\begin{aligned}\alpha &= l_4 - \tfrac{2p_1l_1^2}{a} - \tfrac{3(d_3+|l_2|)^2}{d_1} - \tfrac{3l_3^2}{d_2} \\ M &= \tfrac{3M_1^2}{d_1} + \tfrac{3M_2^2}{d_2} + \tfrac{(M_3+\varepsilon_M)^2}{d_2} + \tfrac{\|K_w\|W_M^2}{2}.\end{aligned} \tag{4.28}$$

According to the above analysis, $\frac{ap_1}{2}, \frac{d_1}{3}, \frac{d_2}{3}, \frac{\|K_w\|}{2}$ are all positive. If (4.22) is satisfied, α can be guaranteed to be positive. Then, from the boundedness analysis in [9], the solution of the estimation error system is uniformly bounded. The bound of the SOC estimation error satisfies $\{|\tilde{x}_1(t)| \leq \sqrt{\frac{M}{p}}\}$, where p satisfies $0 < p < \frac{ap_1}{2}$. If p_1 is selected very large, p can be big enough. Hence, the bound of the SOC estimation error can be arbitrarily small, from which it follows that the RBF neural network-based nonlinear observer is effective for estimating the SOC of the battery.

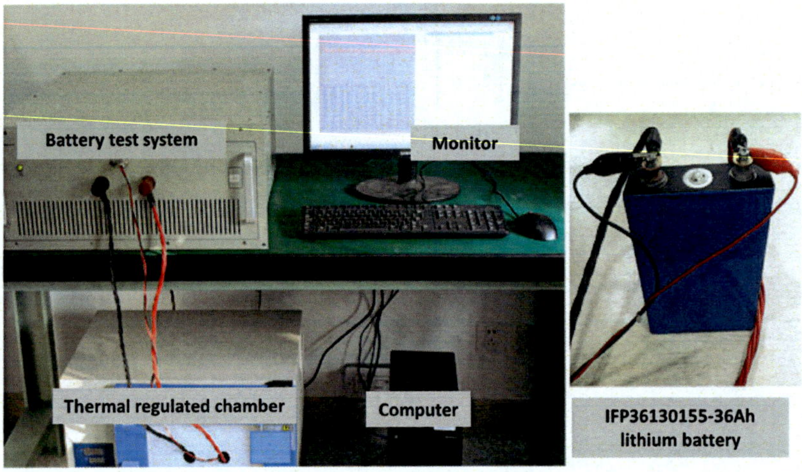

Fig. 4.2 Battery test bench

4.3 Experimental Results

An IFP36130155-36Ah lithium iron phosphate battery is used in the experiments, which has a nominal capacity of 36 Ah, a nominal voltage of 3.7 V, and a cutoff voltage of 2.5 V. The battery and the test bench are illustrated in Fig. 4.2. A thermally regulated chamber is used to ensure that the temperature varies in a small range around 20 °C when the test is performed. The sampling rate of the battery's current and voltage signals is set to 4 Hz. The experimental procedure consists of the following two main parts: a process to extract all parameters in the battery equivalent circuit model and an experiment to verify the performance of the designed neural network-based nonlinear observer.

4.3.1 Experiment for Parameter Extraction

The circuit parameters of the proposed battery model are extracted using a test procedure similar to that of [10]. The battery is discharged with a constant current of 12 A in steps of 5% of the capacity and rests repeatedly for 20 min until its SOC becomes to zero. and Fig. 4.3 shows the voltage response during the rest time period when the SOC is 90 and 85%, where the voltage profile of the battery is smoothed by a low-pass filter. As shown in the Fig. 4.3, the resistance at the corresponding SOCs is calculated as the ratio of the voltage drop U_0 and the current I_B as follows:

$$R_0 = \frac{U_0}{I_B}. \tag{4.29}$$

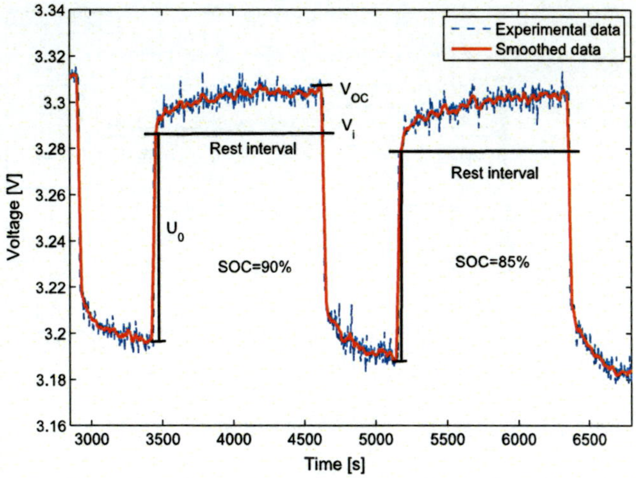

Fig. 4.3 Battery's terminal voltage response and its smoothed value

The parameters of the RC networks at the corresponding measured SOCs can be obtained by fitting the V-I response curve to the following equations that

$$V_B = V_s(1 - e^{-\frac{t}{\tau_s}}) + V_f(1 - e^{-\frac{t}{\tau_f}}) + V_i \tag{4.30}$$

$$V_s = R_s I_B \tag{4.31}$$

$$V_f = R_f I_B = (\frac{(V_{OC} - V_i)}{I_B} - R_s) I_B \tag{4.32}$$

$$\tau_s = R_s C_s \; \tau_f = R_f C_f \tag{4.33}$$

where V_B denotes the terminal voltage of the battery, $I_B = 12$ A is the battery's current before the rest interval, t represents the time during the rest interval, V_s and V_f are the short and long term transient voltage drops with the time constants τ_s and τ_f, and V_i is the battery's voltage after the instantaneous voltage rise. The parameters R_s, C_s, R_f and C_f in the battery model at each measured SOC can be obtained by using the nonlinear least squares method based on (4.30)–(4.33), where V_B, V_{OC}, V_i, t, and I_B are treated as the inputs, τ_s, τ_f, V_s, and V_f are the intermediate variables.

These curves are then fitted with empirical equations consisting of exponential functions and polynomials with a trade-off between accuracy and complexity. Figure 4.4 illustrates the nonlinear relationship between the OCV and SOC of the battery and Fig. 4.5 shows the relationship between the parameters (R_0, R_f, C_f, R_s, and C_s) in the battery equivalent circuit model with respect to the SOC curve. The red dots indicate the experimental data points, and the blue lines indicate the fitted curves corresponding to them. Figure 4.6 compares the simulation results of the men-

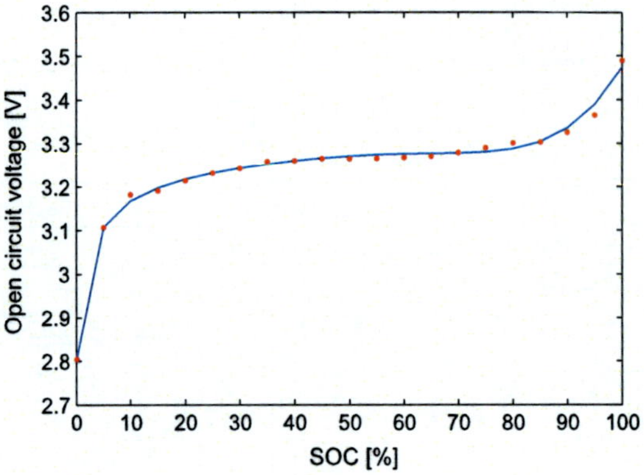

Fig. 4.4 Open circuit voltage versus SOC of the battery

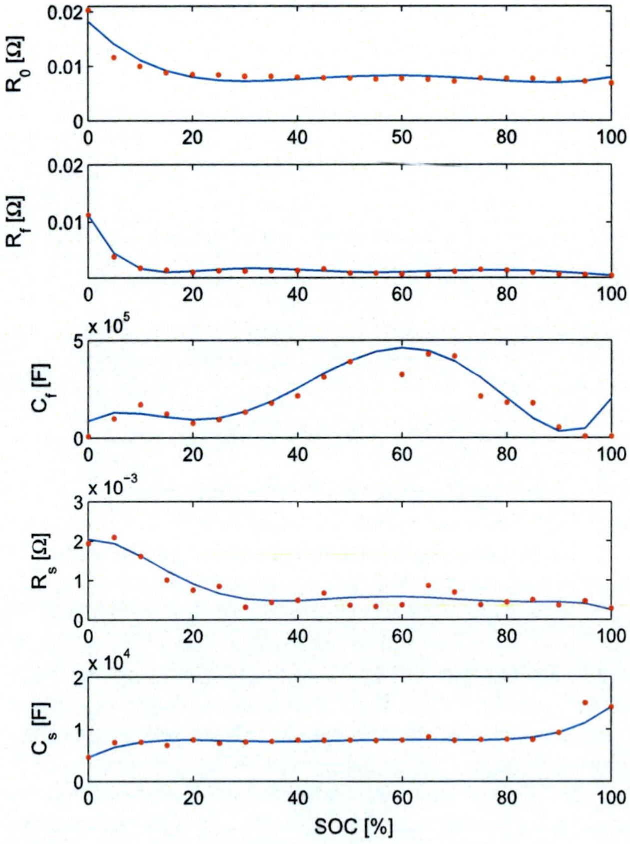

Fig. 4.5 Parameters of the battery's equivalent circuit model

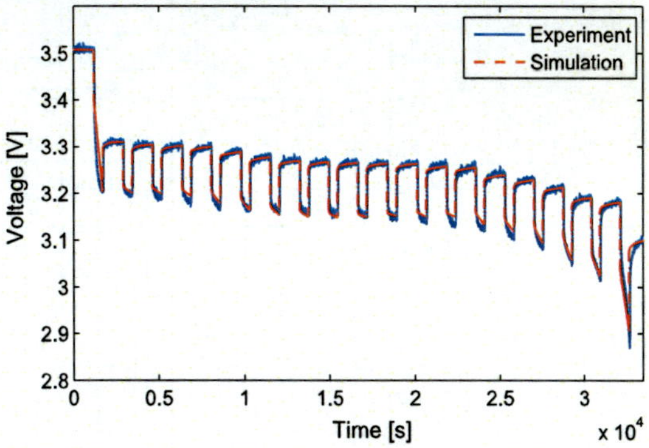

Fig. 4.6 Comparison between simulation results and experimental data

tioned battery equivalent model with the experimental data and it can be seen that the equivalent circuit model matches well with the dynamics of the actual battery.

4.3.2 Experiments for SOC Estimation

The number of neural nodes of the RBF neural network is set to 7. The parameters in the neural network weight vector adaptive law are $K_w = diag\{0.01, 0.01, 0.01, 0.01, 0.01\}$ and $\Gamma = diag\{1, 1, 1, 1, 1, 1, 1\}$. As illustrated in Fig. 4.7, a pulse current is used as the discharge current of the battery, whose amplitude is 12 A, period is 1672 s, and duty ratio is 30%. The observer's initial SOC is randomly selected as 50%. Figure 4.8 shows the uncertainty estimation of the model by the designed RBF neural networks. There is a big error for the step in the pulse current, because the derivative of current is assumed to be 0 in the design of RBF neural network-based nonlinear observer. Figure 4.9 shows the comparison results of the estimated SOC curve by the extended Kalman filter (EKF) and the designed nonlinear observer based on neural network, where the actual SOC is obtained by the ampere-hour counting method. As shown in Fig. 4.10 and Table 4.1, the root mean square (RMS) error of the SOC estimation of the nonlinear observer based on RBF neural network is 1.21%, while that of EKF is 3.06%. It can be seen that the proposed nonlinear observer has higher estimation accuracy than the EKF.

The comparison results of the SOC estimation error are illustrated in Fig. 4.11 for different selected initial estimated SOCs of 50%, 70%, and 90%, respectively. As shown in Table 4.2, the average RMS error of the designed nonlinear observer based on RBF neural network is 1.1233%, indicating that the designed observer has good estimation performance.

The estimation performance and computational burden of the nonlinear observer based on neural network are affected by the number of neurons, and the specific degree of influence is discussed by comparative experiments and simulation. Specif-

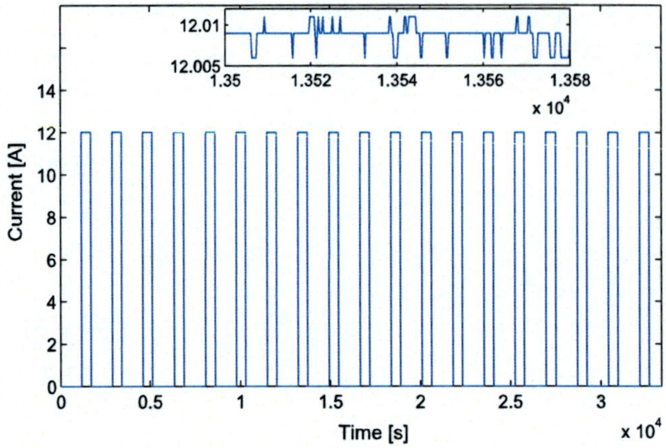

Fig. 4.7 Current profile of the battery

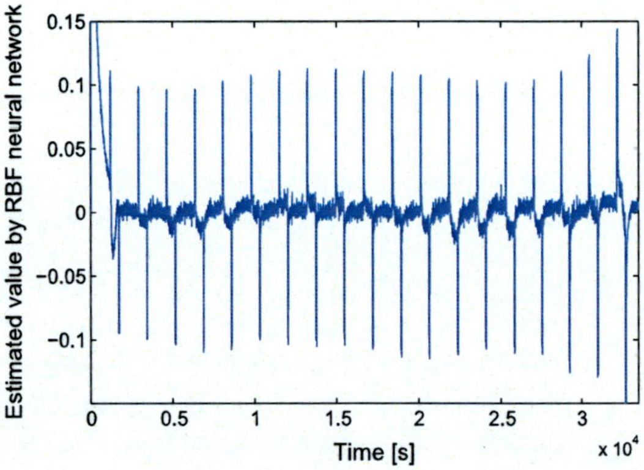

Fig. 4.8 Estimated uncertain term by the RBF neural network

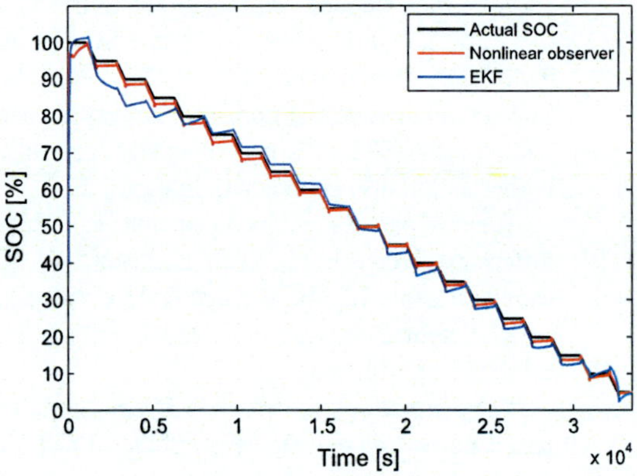

Fig. 4.9 Comparison of the SOC estimated by EKF and the proposed nonlinear observer

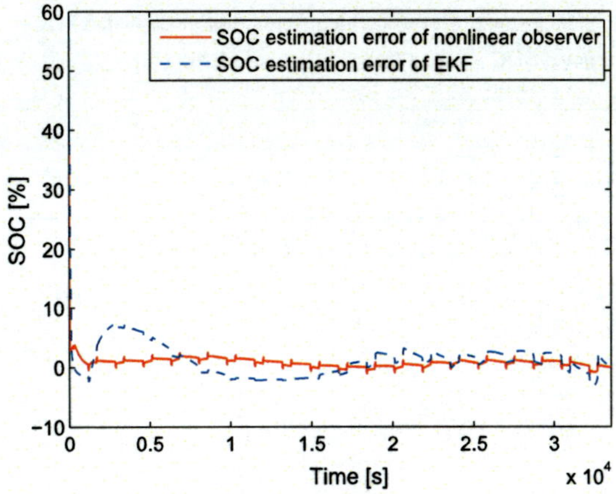

Fig. 4.10 Comparison of SOC estimation error by EKF and the proposed nonlinear observer

Table 4.1 SOC estimation errors with pulse current

	Mean (%)	RMS (%)	Variance
Nonlinear observer	0.88	1.21	6.9189×10^{-5}
EKF	1.16	3.06	8.0197×10^{-4}

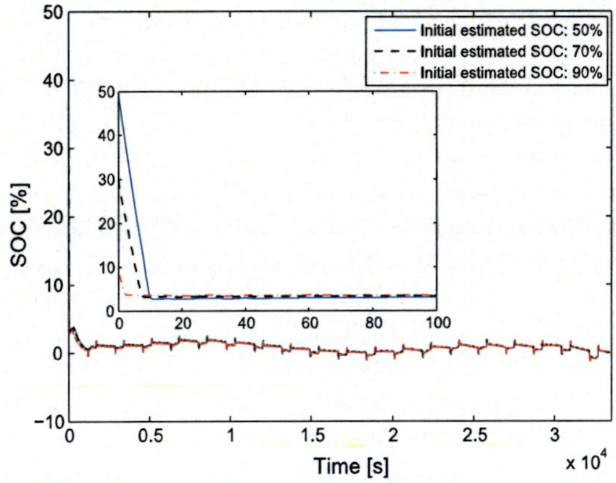

Fig. 4.11 SOC estimation errors of the nonlinear observers with different initial estimated SOCs

Table 4.2 SOC estimation errors with different initial estimated SOCs

Initial estimated SOC (%)	Mean (%)	RMS (%)	Variance
50	0.88	1.21	6.9189×10^{-5}
70	0.86	1.12	5.056×10^{-5}
90	0.82	1.04	4.1265×10^{-5}

Fig. 4.12 SOC estimation errors for the nonlinear observers with different numbers of neural nodes

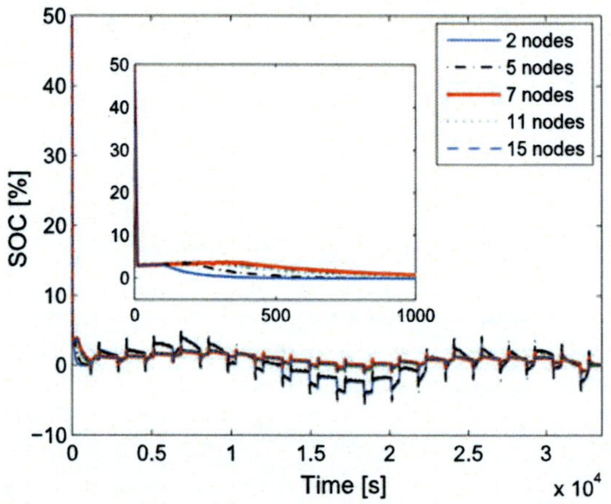

ically, the RBF neural network-based nonlinear observers with 2, 5, 7, 11, and 15 neural nodes are conducted in SIMULINK/MATLAB with the data of battery current and terminal voltage are obtained in the experiment. The comparison of SOC estimation performance is shown in Fig. 4.12, and the computational burden of different neuron numbers is represented by the elapsed times of running the nonlinear observer program in SIMULINK, as shown in Fig. 4.13. From the results, the neural network with seven neural nodes used in our experiments has good performance and a reasonable amount of computation.

Simulations are also conducted in SIMULINK/MATLAB using experimental data to verify the RBF neural network-based nonlinear observer when the battery works under complex current profiles. The federal urban driving schedule (FUDS) is a commonly used driving cycle to evaluate SOC estimation algorithms [11]. Figure 4.14 illustrates the current profile under 20 FUDS test cycles. The initial SOC of the observer is selected as 50%. As illustrated in Fig. 4.15 and Table 4.3, the RMS of the SOC estimation error of the proposed SOC estimation algorithm is 2.23%, which shows that this observer has a faster convergence speed and higher accuracy compared with the EKF.

Fig. 4.13 Elapsed times of the SIMULINK programs of solving the nonlinear observers with different numbers of neural nodes

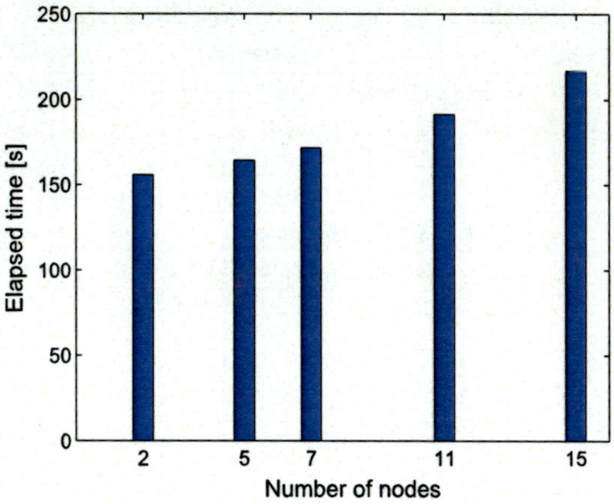

Fig. 4.14 Current profile sampled during 20 consecutive FUDS cycles

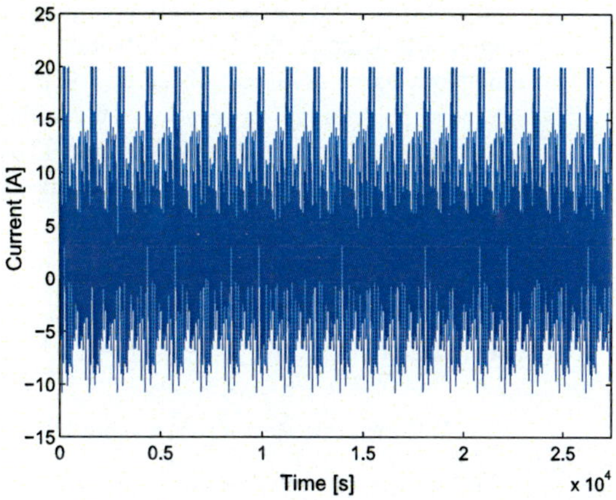

Fig. 4.15 Comparison of SOC estimation errors of EKF and nonlinear observer

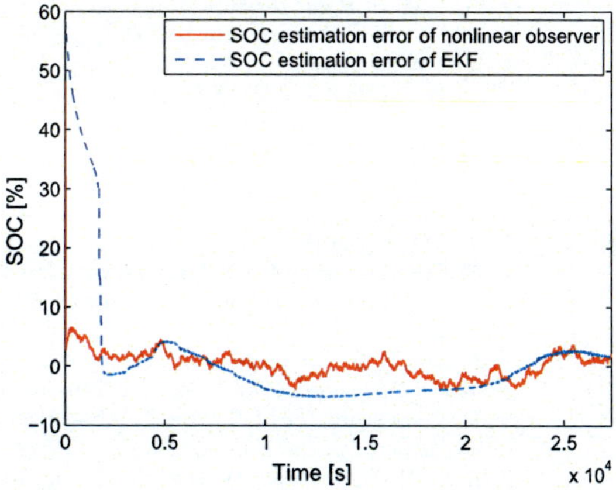

Table 4.3 SOC estimation errors with current of 20 consecutive FUDS cycles

	Mean (%)	RMS (%)	Variance
Nonlinear observer	0.35	2.23	4.8658×10^{-4}
EKF	1.16	10.81	0.0115

Table 4.4 SOC estimation errors for FUDS cycles with different maximum current magnitudes

Maximum current magnitude	RMS (%)	Variance
20 A	2.23	4.8658×10^{-4}
36 A	2.37	4.3397×10^{-4}
50 A	2.48	3.9458×10^{-4}
Average	2.36	4.3838×10^{-4}

Finally, two other experiments composed of 20 consecutive FUDS cycles are conducted to validate the SOC estimation performance of the designed RBF neural network-based nonlinear observer, where the maximum current magnitudes are selected as 36 A and 50 A, respectively. As illustrated in Table 4.4, the average SOC estimation RMS error is 2.36% for these FUDS cycles, which indicates the satisfactory SOC estimation performance of the proposed RBF neural network-based nonlinear observer.

References

1. W. Y. Chang, "The state of charge estimating methods for battery: A review," *ISRN Applied Mathematics*, vol. 2013, pp. 953792, 2013.
2. R. Xiong, J. Cao, Q. Yu, H. He, and F. Sun, "Critical review on the battery state of charge estimation methods for electric vehicles," *IEEE Access*, vol. 6, pp. 1832–1843, 2018.
3. O. Caumont, P. Le Moigne, C. Rombaut, X. Muneret, and P. Lenain, "Energy gauge for lead-acid batteries in electric vehicles," *IEEE Transactions on Energy Conversion*, vol. 15, no. 3, pp. 354–360, 2000.
4. S. Lee, J. Kim, J. Lee, and B.H. Cho, "State-of-charge and capacity estimation of lithium-ion battery using a new open-circuit voltage versus state-of-charge," *Journal of Power Sources*, vol. 185, no. 2, pp. 1367–1373, 2008.
5. J. Chen, Q. Ouyang, C. Xu, and H. Su, "Neural network-based state of charge observer design for lithium-ion batteries," *IEEE Transactions on Control Systems Technology*, vol. 26, no. 1, pp. 313–320, 2018.
6. M. Chen, A. Rincon-Mora, "Accurate electrical battery model capable of predicting runtime and I-V performance," *IEEE Transactions on Energy Conversion*, vol. 21, no. 2, pp. 504–511, 2006.
7. J. Park, I. W. Sandberg, "Universal approximation using radial-basis-function networks," *Neural Computation*, vol. 3, no. 2, pp. 246–257, 1991.
8. M. S. de Queiroz, D. M. Dawson, S. P. Nagarkatti, and F. Zhang, *Lyapunov-Based Control of Mechanical Systems*. Boston, MA, USA: Birkhäuser, 2000.
9. H. K. Khalil, *Nonlinear Systems*, 3rd ed. Englewood Cliffs, NJ, USA: Prentice-Hall, 2002.

10. S. Abu-Sharkh, D. Doerffel, "Rapid test and non-linear model characterisation of solid-state lithium-ion batteries," *Journal of Power Sources*, vol. 130, nos. 1–2, pp. 266–274, 2004.
11. H. He, R. Xiong, X. Zhang, F. Sun, and J. Fan, "State-of-charge estimation of the lithium-ion battery using an adaptive extended Kalman filter based on an improved Thevenin model," *IEEE Transactions on Vehicular Technology*, vol. 60, no. 4, pp. 1461–1469, 2011.

Chapter 5
Co-estimation of State of Charge and Model Parameters for Lithium-Ion Batteries

The SOC estimation algorithm mentioned in Chap. 4 is based on a battery model with parameters explicitly known in advance. It means that there must be a parameter identification procedure to prepare the battery's model for SOC estimation. But this parameter identification procedure can be tedious and even needs to be run repeatedly to extract the correct parameters for an aging battery. Hence, it raises an important question: how to design an SOC estimation method when the battery's model parameters are initially unknown. In addition, it is common to encounter outliers, i.e., unrepresentative data points that deviate significantly from normal values, during battery signal measurements. These outliers may come from measurement failures or big noise disturbances. In practice, they can degrade the SOC estimation performance by introducing bias and, in some extreme cases, can even trigger a complete failure of estimation.

In this chapter, a robust recursive least squares (RLS) algorithm is used to extract the parameters of the equivalent circuit model of lithium-ion batteries on-line. Compared with the conventional RLS algorithm, superior parameter identification performance of this strategy can be achieved in spite of outliers in battery measurement signals. Then, based on the model with identified parameters, a robust observer is proposed to estimate the battery's SOC, which can effectively suppress the disturbance brought by unknown model errors [1].

5.1 Battery Model

The equivalent circuit model shown in Fig. 5.1 is used to describe the dynamic characteristics of the battery because it strikes a excellent balance between model accuracy and structural simplicity. As shown in Fig. 5.1, C_b is a capacitor used to store the charge, R_0 is the internal resistance, and the RC circuit (R_t, C_t) can represent the voltage transients during charging/discharging. In general, assuming that the battery

Q. Ouyang and J. Chen, *Advanced Model-Based Charging Control for Lithium-Ion Batteries*, https://doi.org/10.1007/978-981-19-7059-7_5

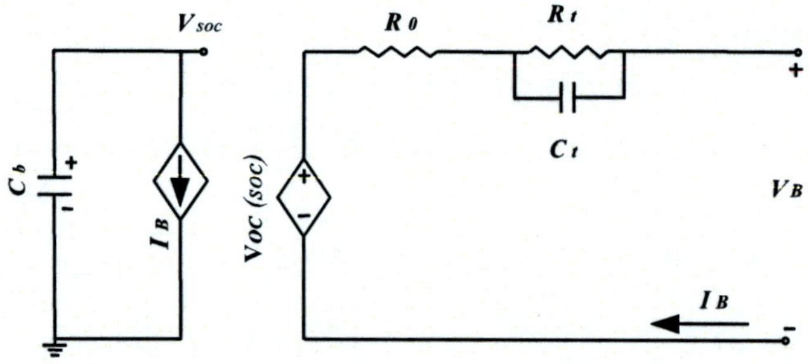

Fig. 5.1 Equivalent circuit model of the battery

is fully charged (100% SOC) when the voltage across C_b is 1 V and fully discharged (0% SOC) when the voltage is 0 V. The mapping from the SOC to the OCV can be described as follows:

$$V_{OC} = g(SOC) \tag{5.1}$$

where V_{OC} and SOC represents the OCV and SOC of the battery, $g(\cdot)$ is a nonlinear monotonically increasing function. Assuming that the battery's current keeps constant during each sampling interval, the battery model can be generated as:

$$\begin{aligned} &SOC(k+1) = SOC(k) - \tfrac{\eta_0 T}{C_b} I_B(k) \\ &]V_t(k+1) = e^{-\frac{T}{R_t C_t}} V_t(k) + (1 - e^{-\frac{T}{R_t C_t}}) R_t I_B(k) \\ &]V_B(k) = g(SOC(k)) - R_0 I_B(k) - V_t(k) \end{aligned} \tag{5.2}$$

where $V_t(k)$ is the voltage across the capacitor C_t, $I_B(k)$ represents the battery's current, $V_B(k)$ is the terminal voltage, η_0 denotes the Coulomb coefficient, and T is the sampling period. In practice, there exists model noise and uncertainty in the equivalent circuit model. With considering them and defining the state $x(k) \in \mathbb{R}^2$ and the parameter vector $\theta \in \mathbb{R}^3$ as

$$\begin{aligned} x(k) &= [x_1(k), x_2(k)]^T \triangleq [SOC(k),\ V_t(k)]^T \\ \theta &= [\theta_1,\ \theta_2,\ \theta_3]^T \triangleq [e^{-\frac{T}{R_t C_t}},\ R_t,\ R_0]^T \end{aligned} \tag{5.3}$$

the battery model (5.2) can be rewritten to the following state-space form

$$\begin{aligned} x(k+1) &= A(\theta)x(k) + B(\theta)\xi(k) + w(k) \\ y(k) &= h(x(k), \theta, \xi(k)) + v(k) \end{aligned} \tag{5.4}$$

where the output is $y(k) \triangleq V_B(k) \in \mathbb{R}$; $\xi(k) \triangleq I_B(k) \in \mathbb{R}$ can be considered as measurable parameters; $w(k)$ and $v(k)$ are unknown model deviations or noise, and they can be assumed to be bounded; the matrices $A(\theta) \in \mathbb{R}^{2\times 2}$, $B(\theta) \in \mathbb{R}^2$, and $h(\cdot) \in \mathbb{R}$ in (5.4) are

$$A(\theta) = \begin{bmatrix} 1 & 0 \\ 0 & \theta_1 \end{bmatrix}, \quad B(\theta) = \begin{bmatrix} -d \\ (1-\theta_1)\theta_2 \end{bmatrix}$$

$$h(x(k), \theta, \xi(k)) = g(x_1(k)) - x_2(k) - \theta_3 \xi(k)$$

with $d = \frac{\eta_0 T}{C_b}$. The nonlinear relationship between the OCV and SOC of the battery can be obtained by directly measuring the terminal voltage of the battery at the corresponding SOC after the battery has been at rest for a long enough time. But the parameters θ_i $(1 \leq i \leq 3)$ in the equivalent circuit model must first be determined by complex prior experiments followed by fitting methods. In practice, these parameters change slowly rather than constant with many factors, such as the SOC, temperature, and operating conditions of the battery. Therefore, it is of great importance to propose a method for online identification of θ in (5.4), which avoids complex prior model parameter extraction experiments while maintaining sufficient accuracy of the battery model.

5.2 Co-estimation of Model Parameters and SOC

Based on the battery's discrete state-space model (5.4) developed in the above section, a robust strategy will be proposed here to estimate the model parameters and SOC of the battery simultaneously, which can be divided into two parts. A robust RLS approach is developed for battery model parameter on-line identification and a robust observer is designed to accurately estimate the SOC of the battery based on the identified model.

5.2.1 On-line Battery Model Parameter Identification

Since the value of C_b is much greater than that of $I_B(k)$, it is known from the dynamics of the battery model that the variation of the SOC and OCV of the battery is small enough in each sampling interval, which can be deduced that $g(x_1(k)) - g(x_1(k-1)) = 0$. Then, with defining $y_1(k) = y(k) - y(k-1)$ and $\xi_1(k) = \xi(k) - \xi(k-1)$, it can be calculated from (5.4) that

$$\begin{aligned} y_1(k) &= \theta_1 y_1(k-1) - \theta_3 \xi_1(k) - ((1-\theta_1)\theta_2 \\ &\quad - \theta_1\theta_3)\xi_1(k-1) + v_1(k) \\ &= U^T(k)\Phi + v_1(k) \end{aligned} \tag{5.5}$$

with

$$\begin{aligned} \Phi &\triangleq [\theta_1, -\theta_3, \theta_1\theta_3 - (1-\theta_1)\theta_2]^T \\ U(k) &\triangleq [y_1(k-1), \xi_1(k), \xi_1(k-1)]^T \end{aligned} \tag{5.6}$$

where $v_1(k)$ denotes the unknown model noise or bias (including the model bias due to the above assumption about the battery's OCV). It can be assumed to be bounded and satisfy

$$|v_1(k)| \leq v_{1M} \tag{5.7}$$

where v_{1M} is a positive constant.

From the least squares (LS) theory [2], the estimation of the parameter vector $\Phi(k)$ can be obtained through

$$\hat{\Phi}(k) = \arg\min_{\Phi} \sum_{i=1}^{k} \lambda^{k-i} \rho(\tilde{y}_1(i|\Phi)) \tag{5.8}$$

with

$$\tilde{y}_1(k|\Phi) = y_1(k) - U^T(k)\Phi$$

where $\hat{\Phi}(k) = [\hat{\phi}_1(k), \hat{\phi}_2(k), \hat{\phi}_3(k)]^T \triangleq [\hat{\theta}_1(k), \hat{\theta}_3(k), \hat{\theta}_2(k) - \hat{\theta}_1(k)\hat{\theta}_3(k)]^T \in \mathbb{R}^3$ is the estimation of Φ; λ is the forgetting factor, which is the weight of the prediction error on a moment in time, usually chosen between 0.98 and 0.995; $\rho(\tilde{y}_1)$ denotes the loss function, $\rho(\tilde{y}_1) = \frac{1}{2}\tilde{y}_1^2$, where the conventional LS criterion chooses to minimize the data fitting error, as suggested by the conventional LS strategy [3].

The traditional LS criterion is appropriate when the prediction errors can be assumed to be independently Gaussian distributed errors with mean 0. It is important to note that in practice, such as in electric vehicle applications, outliers of some battery measurement signals are often encountered, which seriously violate the Gaussian assumption, and thus estimation based on the traditional LS criterion may affect the parameter identification of the battery model and reduce the accuracy of the identification. Therefore, a robust strategy is needed to solve this problem to make the model parameter identification less sensitive to the outliers of the measurement data. A robust criterion function [3] on the loss function is set, which assigns a limited weight to the anomalous estimation errors and reduces the sensitivity of the estimation procedure to large errors, which is defined as

$$\rho(\tilde{y}_1) = \begin{cases} \frac{1}{2}\tilde{y}_1^2, & \text{if} \quad |\tilde{y}_1| \leq \alpha\delta \\ \frac{1}{2}(\alpha\delta)^2, & \text{if} \quad |\tilde{y}_1| > \alpha\delta \end{cases} \tag{5.9}$$

where δ denotes the standard deviation of $\tilde{y}_1$ and α is a user-definable positive constant. the smaller α is, the smaller the second term in the above equation is, and when there is a large estimation error, the impact of this error on the system estimation is limited, so the system is more robust. However, for data without outliers will reduce the efficiency of the algorithm, so α needs to be defined according to the user's needs to achieve a good balance of system robustness and algorithm efficiency. The estimation error is assumed to be a Gaussian distribution with contamination, i.e., a Gaussian distribution disturbed by a few outliers. Based on the 3-$sigma$ hard rejection rule [4] it is known that if a set of data obeys normal distribution, δ denotes the standard deviation of the data 99.7% of the estimation error $\tilde{y}_1$ falls into the 3-$sigma$ range and can be regarded as normal, and the probability that the estimation error $\tilde{y}_1$ falls outside the range of 3-$sigma$ is only 0.3%, which can be considered as an outlier and assigned a limited weight to it. Therefore, $\alpha = 3$ is used in this paper. As shown in (5.8 and 5.9), the impact of the estimation error on $\hat{\Phi}$ can be confined to $\frac{1}{2}(\alpha\delta)^2$ when a large estimation errors $\tilde{y}_1$ occur due to outliers in the measured data. By introducing a robustness criterion function on the loss function, the error in model parameter identification caused by outliers can be effectively reduced and the accuracy of parameter identification can be improved. In practice, the standard deviation δ of the estimation error is unknown. Here the common practice of using the median filter to obtain the absolute median difference [4] is used to estimate the standard deviation δ as follows

$$\hat{\delta} = \frac{\text{median}\{|\tilde{y}_1(i) - \text{median}(\tilde{y}_1(i))|,\ i = k - N_m, \ldots, k\}}{0.6745} \tag{5.10}$$

where N_m is the length of the selected data. There is no general closed form solution for the robust criterion function defined by previous equation. Further, based on the robust criterion function (5.9), a robust RLS algorithm for on-line parameter identification of the battery model is designed as

$$\hat{\Phi}(k) = \hat{\Phi}(k-1) + \Gamma(k)\dot{\rho}(\tilde{y}_1(k|\hat{\Phi}(k-1))) \tag{5.11}$$

with

$$\begin{aligned}
\tilde{y}_1(k|\hat{\Phi}(k-1)) &= y_1(k) - U^T(k)\hat{\Phi}(k-1)\\
\Gamma(k) &= \frac{P(k-1)U(k)}{\lambda + \ddot{\rho}(\tilde{y}_1(k|\hat{\Phi}(k-1)))U^T(k)P(k-1)U(k)}\\
P(k) &= \frac{P(k-1) - \Gamma(k)U^T(k)P(k-1)\ddot{\rho}(\tilde{y}_1(k|\hat{\Phi}(k-1)))}{\lambda}
\end{aligned}$$

where $\Gamma(k) \in \mathbb{R}^3$ is the gain vector, $P(k) \in \mathbb{R}^{3\times3}$ recursively summarizes the parameter identification error before step k, $\dot{\rho}(\cdot)$ and $\ddot{\rho}(\cdot)$ are the first and second derivatives of $\rho(\cdot)$ in (5.9), which can be deduced that

$$\dot{\rho}(\tilde{y}_1) = \begin{cases} \tilde{y}_1, & \text{if } |\tilde{y}_1| \le \alpha\delta \\ 0, & \text{if } |\tilde{y}_1| > \alpha\delta \end{cases}$$

$$\ddot{\rho}(\tilde{y}_1) = \begin{cases} 1, & \text{if } |\tilde{y}_1| \le \alpha\delta \\ 0, & \text{if } |\tilde{y}_1| > \alpha\delta. \end{cases}$$

Convergence analysis: As mentioned above, the estimated parameters of the designed robust RLS algorithm (5.11) remain unchanged as $\hat{\Phi}(k) = \hat{\Phi}(k-1)$ with $P(k) = \frac{1}{\lambda}P(k-1) \approx P(k-1)$ when there are measurement outliers, which does not affect the convergence of the estimated model parameters. Therefore, a convergence proof is next provided for the battery measurement signal without outliers. For normal signals,$\dot{\rho}(\tilde{y}_1) = \tilde{y}_1$ and $\ddot{\rho}(\tilde{y}_1) = 1$, the robust RLS (ref2.54) follows the same procedure as the conventional RLS algorithm. Then, it can be inferred that

$$P(k) = \tfrac{1}{\lambda}(I - P(k-1)U(k)aU^T(k))P(k-1) \tag{5.12}$$

with

$$a = (\lambda + U^T(k)P(k-1)U(k))^{-1}$$

where I denotes the identity matrix with appropriate dimensions. Using the matrix inversion formula $A^{-1} - A^{-1}B(D + CA^{-1}B)^{-1}CA^{-1} = (A + BD^{-1}C)^{-1}$, (5.12) can be written as

$$P^{-1}(k) = \lambda P^{-1}(k-1) + U(k)U^T(k). \tag{5.13}$$

Under the assumption that $U(k)$ is continuously excited, $P(k)$ can be a bounded positive definite matrix if the initial value of $P(k)$ is chosen to be $P(0) = P^T(0) > 0$, [5]. This continuous excitation condition can be satisfied in many applications, e.g., the battery current varies frequently when the electric vehicle is operating. Defining the parameter estimation error as

$$\tilde{\Phi}(k) \triangleq \Phi - \hat{\Phi}(k). \tag{5.14}$$

It can be deduced from (5.11) that

$$\begin{aligned}\tilde{\Phi}(k) = &(I - aP(k-1)U(k)U^T(k))\tilde{\Phi}(k-1)\\ &- aP(k-1)U(k)v_1(k).\end{aligned} \tag{5.15}$$

A Lyapunov candidate function is chosen as $V(k) = \tilde{\Phi}^T(k)P^{-1}(k)\tilde{\Phi}(k) \in \mathbb{R}$. According to (5.11) and (5.13), we can be obtained that $P^{-1}(k-1) \geq \lambda^{k-1}P^{-1}(0)$, and the change of the Lyapunov function $\Delta V(k) = V(k) - V(k-1)$ is calculated as

$$\begin{aligned}\Delta V(k) = &-\tilde{\Phi}^T(k-1)((1-\lambda)P^{-1}(k-1) + a\lambda U(k)\\ &\times U^T(k))\tilde{\Phi}(k-1) - 2a\lambda\tilde{\Phi}^T(k-1)U(k)v_1(k)\\ &+ av_1^T(k)U^T(k)P(k-1)U(k)v_1(k)\\ \leq &-(1-\lambda)\tilde{\Phi}^T(k-1)P^{-1}(k-1)\tilde{\Phi}(k-1) + v_{1M}^2\\ \leq &-(1-\lambda)\lambda^{k-1}\lambda_m(P^{-1}(0))\tilde{\Phi}^T(k-1)\tilde{\Phi}(k-1) + v_{1M}^2\end{aligned} \tag{5.16}$$

where $\lambda_m(P^{-1}(0))$ denotes the minimum eigenvalue of $P^{-1}(0)$. Since $0 < \lambda < 1$ and $P(0)$ is positive definite, it can be obtained that $(1-\lambda)\lambda^{k-1}\lambda_m(P^{-1}(0))$ is a positive value. From (5.16), $\Delta V(k)$ is negative outside the set $S \triangleq \{||\tilde{\Phi}(k-1)|| \leq \frac{v_{1M}^2}{(1-\lambda)\lambda^{k-1}\lambda_m(P^{-1}(0))}\}$, which demonstrates that the Lyapunov function $V(k)$ monotonically decreases until the parameter estimation error $\tilde{\Phi}$ enters into the set S and cannot leave this set from that time on. With $P(0)$ selected to make $\lambda_m(P^{-1}(0))$ large enough, S can be a small range. Based on the boundedness analysis[6], the parameter estimation error is uniformly bounded. Note that for the case of the exact battery model with $v_1(k) \approx 0$, the parameter estimation error can converge to almost zero from (5.16).

By (5.11) estimating $\Phi(k)$ accurately, the battery model parameter θ can be determined as

$$\hat{\theta}(k) = [\hat{\phi}_1(k), \tfrac{-\hat{\phi}_1(k)\hat{\phi}_2(k)-\hat{\phi}_3(k)}{1-\hat{\phi}_1(k)}, -\hat{\phi}_2(k)]^T \tag{5.17}$$

where $\hat{\theta}(k)$ denotes the estimated battery model parameters, which will be used in the robust observer designed below for SOC estimation.

5.2.2 *Robust Observer for SOC Estimation*

Using the on-line identified parameter $\hat{\theta}(k)$ in (5.17) to replace θ, the battery model (5.4) can be expressed as

$$\begin{aligned} x(k+1) &= A(\hat{\theta}(k))x(k) + B(\hat{\theta}(k))\xi(k) + w_2(k) \\ y(k) &= h(x(k), \hat{\theta}(k), \xi(k)) + v_2(k) \end{aligned} \tag{5.18}$$

with

$$\begin{aligned} w_2(k) &= \Delta A(k)x(k) + \Delta B(k)\xi(k) + w(k) \\ v_2(k) &= \Delta h(k) + v(k) \end{aligned}$$

where

$$\begin{aligned} \Delta A(k) &= A(\theta) - A(\hat{\theta}(k)) \\ \Delta B(k) &= B(\theta) - B(\hat{\theta}(k)) \\ \Delta h(k) &= h(x(k), \theta, \xi(k)) - h(x(k), \hat{\theta}(k), \xi(k)). \end{aligned}$$

$\Delta A(k)$, $\Delta B(k)$, and $\Delta h(k)$ are considered as unknown model noise due to parameter misidentification. $w_2(k)$ and $v_2(k)$ are assumed to be bounded. To estimate the state $x(k)$ of the battery, a robust observer is proposed as

$$\begin{aligned} \hat{x}(k+1) = {} & A(\hat{\theta}(k))\hat{x}(k) + B(\hat{\theta}(k))\xi(k) \\ & + L(y(k) - h(\hat{x}(k), \hat{\theta}(k), \xi(k))) \end{aligned} \tag{5.19}$$

where $\hat{x}(k) = [\hat{x}_1(k),\ \hat{x}_2(k)]^T \triangleq [S\hat{O}C(k), \hat{V}_t(k)]^T \in \mathbb{R}^2$ is the estimation of $x(k)$, and $L \in \mathbb{R}^2$ is the gain vector to be designed. Using Taylor expansion, it yields

$$g(x_1(k)) = g(\hat{x}_1(k)) + \tfrac{\mathrm{d}g(\hat{x}_1(k))}{\mathrm{d}x_1}\tilde{x}_1(k) + R_n(\tilde{x}_1(k)) \tag{5.20}$$

where $\tilde{x}(k) \triangleq x(k) - \hat{x}(k) \in \mathbb{R}^2$ is defined as the state estimation error, $R_n(\tilde{x}_1(k))$ is the high-order term. From (5.18) and (5.20), it can be obtained that

$$y(k) - h(\hat{x}(k), \hat{\theta}(k), \xi(k)) = C(k)\tilde{x}(k) + v_3(k) \tag{5.21}$$

with

$$C(k) = [\tfrac{dg(\hat{x}_1(k))}{x_1(k)}, -1]$$
$$v_3(k) = R_n(\tilde{x}_1(k)) + v_2(k).$$

Then, based on (5.18)–(5.21), the state estimation error system can be rewritten as

$$\tilde{x}(k+1) = (A(\hat{\theta}(k)) - LC(k))\tilde{x}(k) + (E - LF)\bar{w}(k) \tag{5.22}$$

with $E = [I,\ 0_2] \in \mathbb{R}^{2\times3}$, $F = [0_2^T,\ 1] \in \mathbb{R}^{1\times3}$, $\bar{w}(k) = [w_2^T(k),\ v_3(k)]^T \in \mathbb{R}^3$, where 0_2 is a zero column vector.

Convergence analysis: To analyze the convergence of the proposed robust observer, the following Theorem is proposed.

Theorem 5.1 *The battery state estimation error system (5.22) is stable for a given attenuation $\gamma > 0$, if there exists a positive definite matrix $P_1 = P_1^T \in \mathbb{R}^{2\times2}$ and the gain vector $L = P_1^{-1}Q$ satisfying the following linear matrix inequality (LMI):*

$$\begin{bmatrix} -P_1 & \Pi_1 & \Pi_2 \\ \Pi_1^T & I - P_1 & 0_{2\times3} \\ \Pi_2^T & 0_{3\times2} & -\gamma^2 I \end{bmatrix} < 0 \tag{5.23}$$

with

$$\Pi_1 = P_1 A_M - QC(k)$$
$$\Pi_2 = P_1 E - QF$$

where A_M is selected as $A_M \geq A(\hat{\theta}(k))$.

Proof A Lyapunov function is chosen as $V_2(k) = \tilde{x}^T(k)P_1\tilde{x}(k)$, and the change of this function can be calculated as

$$\begin{aligned} \Delta V_2(k) &= V_2(k+1) - V_2(k) \\ &= \begin{bmatrix} \tilde{x}(k) \\ \bar{w}(k) \end{bmatrix}^T \begin{bmatrix} \Omega_{11} & \Omega_{12} \\ \Omega_{12}^T & \Omega_{22} \end{bmatrix} \begin{bmatrix} \tilde{x}(k) \\ \bar{w}(k) \end{bmatrix} \end{aligned} \tag{5.24}$$

with

$$\Omega_{11} = (A(\hat{\theta}(k)) - LC(k))^T P_1 (A(\hat{\theta}(k)) - LC(k)) - P_1$$
$$\Omega_{12} = (A(\hat{\theta}(k)) - LC(k))^T P_1 (E - LF)$$
$$\Omega_{22} = (E - LF)^T P_1 (E - LF).$$

From (5.24), it can be obtained that

$$\begin{aligned}&\Delta V_2(k) + \tilde{x}^T(k)\tilde{x}(k) - \gamma^2\bar{w}^T(k)\bar{w}(k)\\&= \begin{bmatrix}\tilde{x}(k)\\ \bar{w}(k)\end{bmatrix}^T \begin{bmatrix}\Omega_{11}+I & \Omega_{12}\\ \Omega_{12}^T & \Omega_{22}-\gamma^2 I\end{bmatrix} \begin{bmatrix}\tilde{x}(k)\\ \bar{w}(k)\end{bmatrix}.\end{aligned} \tag{5.25}$$

With (5.23), $P_1 > 0$, and $A_M \geq A(\hat{\theta}(k))$ satisfied, it yields that

$$\begin{bmatrix}-P_1 & \Pi_3 & \Pi_2\\ \Pi_3^T & I-P_1 & 0_{2\times 3}\\ \Pi_2^T & 0_{3\times 2} & -\gamma^2 I\end{bmatrix} < 0 \tag{5.26}$$

with

$$\Pi_3 = P_1 A(\hat{\theta}(k)) - QC(k).$$

According to Schur complement [7], (5.26) is equivalent to

$$\begin{bmatrix}\Omega_{11}+I & \Omega_{12}\\ \Omega_{12}^T & \Omega_{22}-\gamma^2 I\end{bmatrix} < 0. \tag{5.27}$$

Hence, based on (5.25) and (5.27), it yields

$$\tilde{x}^T(k)\tilde{x}(k) - \gamma^2\bar{w}^T(k)\bar{w}(k) < -\Delta V_2(k),\ \forall k \geq 0. \tag{5.28}$$

From (5.28), for $\forall N \geq 0$, it can be deduced that

$$\begin{aligned}\sum_{k=0}^{N}\tilde{x}^T(k)\tilde{x}(k) &< V_2(0) - V_2(N) + \gamma^2\sum_{k=0}^{N}\bar{w}^T(k)\bar{w}(k)\\ &< \tilde{x}^T(0)P_1\tilde{x}(0) + \gamma^2\sum_{k=0}^{N}\bar{w}^T(k)\bar{w}(k).\end{aligned} \tag{5.29}$$

Therefore, from (5.29) and referring to [8], a convergence proof for the battery state estimation error system has been achieved. For the cases where the SOC estimation has lasted long enough, the effect of the initial condition $\tilde{x}(0)$ can be assumed to be zero, and the estimation performance can be replaced by $||\tilde{x}(k)|| \leq \gamma||\bar{w}(k)||$ [9].

As shown in Theorem 5.1, γ plays an important role in the feasibility of (5.23) to choose the suitable gain L. Based on the robust control theory [10], the following optimization issue can be formulated for the optimal γ selection

$$\begin{aligned}&\min \gamma^2\\ &\text{s.t.}\\ &\quad \begin{bmatrix}-P_1 & \Pi_1 & \Pi_2\\ \Pi_1^T & I-P_1 & 0_{2\times 3}\\ \Pi_2^T & 0_{3\times 2} & -\gamma^2 I\end{bmatrix} < 0\\ &\quad P_1 > 0\end{aligned} \tag{5.30}$$

where P_1 and Q are the optimization variables. The optimal gain vector in the designed robust observer can be obtained as $L = P_1^{-1} Q$. Note that the LMI-based optimization problem (5.30) can be easily solved by the LMI toolbox in MATLAB.

5.2.3 Summary of the Overall SOC Estimation Strategy

From the above-mentioned convergence analysis, the proposed robust RLS needs sufficient current excitation to achieve stable model identification. It can be satisfied in many cases, such as the driving cycles of electric vehicles. However, challenges remain in some scenarios, such as long periods of rest and constant current charging/discharging. In these situations, the performance of the model parameter estimation degrades. To address this issue, we design an additional loop where the robust RLS is interrupted and the average of the historically estimated model parameters is utilized when there exists low-current excitation as follows:

$$\hat{\theta}(k) = \begin{cases} [\hat{\phi}_1(k), \frac{-\hat{\phi}_1(k)\hat{\phi}_2(k)-\hat{\phi}_3(k)}{1-\hat{\phi}_1(k)}, -\hat{\phi}_2(k)]^T, \text{ if } T_n < T_a \\ \text{avrg}\{\hat{\theta}(i), i = k-1, \ldots, k-n+1\}, \text{ if } T_n \geq T_a \end{cases} \tag{5.31}$$

where T_n is the number of steps for $I_B(k) = \cdots = I_B(k-n+1)$ and T_a is the specified period to determine whether the battery is in the low-current excitation case, which in practice is recommended to be chosen to be a few tens of seconds. By using (5.31), the estimated model parameters remain almost constant in the low-current excitation case and start to converge to the actual values in the high-current excitation case. Since the battery model parameters are slowly changing variables, the model parameter estimation errors always remain within a certain range in the low-current excitation case, which does not affect the convergence of the model parameter estimation in the high-current excitation case. Therefore, the designed model parameter estimation method is feasible for various operating conditions including low-current excitation.

5.3 Experimental Results

Experiments are conducted in this section to validate the estimation method designed above with the battery test bench illustrated in Fig. 5.2. An IFP36130155-36Ah lithium iron phosphate battery is used with a nominal capacity of 36 Ah, a nominal voltage of 3.2 V, and a cutoff voltage of 2.5 V. The relationship between the battery's OCV and SOC is depicted in Fig. 5.3. The sampling rates of the current and voltage signals are set to 1 Hz. The current of the battery is shown in Fig. 5.4. To test the performance of the proposed SOC estimation algorithm in the presence of unexpected outliers of the battery measurement signal in practice, outliers with the length of 1 s

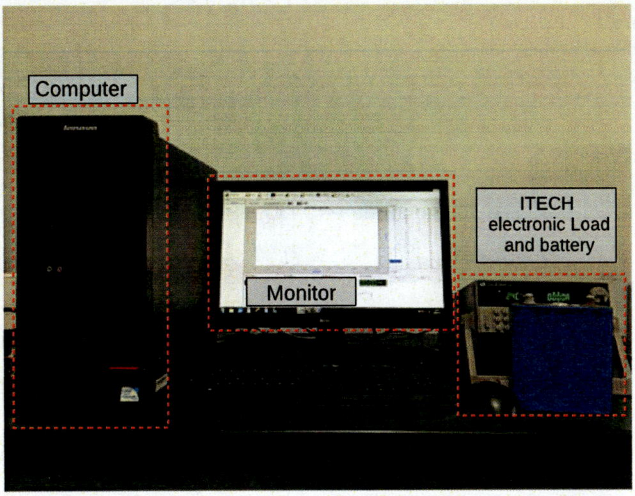

Fig. 5.2 Battery test bench

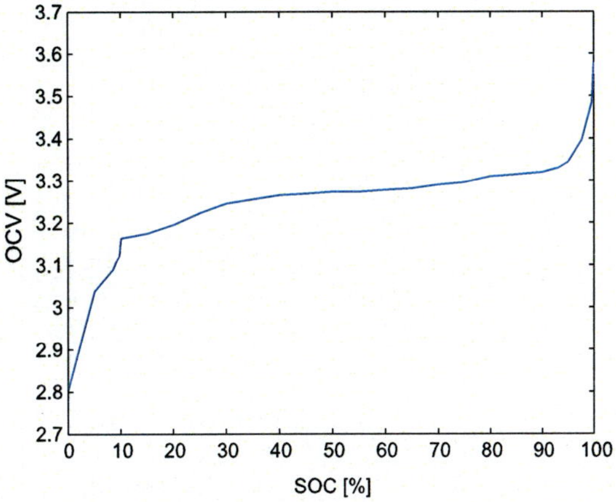

Fig. 5.3 Mapping from the battery's SOC to OCV

and amplitude of 0.3 V are randomly selected and artificially injected in the original terminal voltage signals of 10000 s and 20000 s, respectively.

Offline calculation of model parameters: From (5.5), the model parameters calculated offline using the LS method can be obtained as

$$\hat{\Phi} = (U^T U)^{-1} U^T Y \tag{5.32}$$

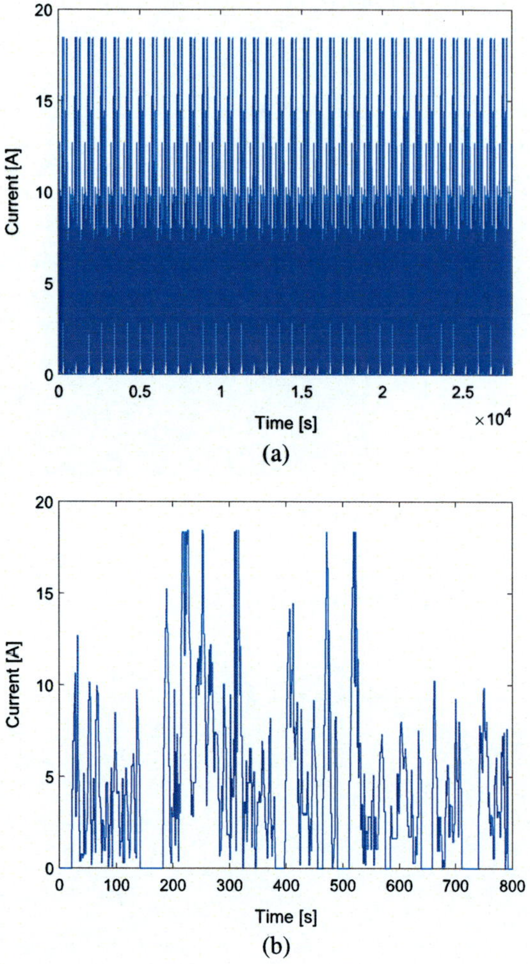

Fig. 5.4 **a** Applied current profile, **b** zoom of the applied current profile

with

$$Y = \begin{bmatrix} y_1(2) \\ y_1(3) \\ \cdots \\ y_1(N) \end{bmatrix}, \quad U = \begin{bmatrix} y_1(1) & \xi_1(2) & \xi_1(1) \\ y_1(2) & \xi_1(3) & \xi_1(2) \\ \cdots & \cdots & \cdots \\ y_1(N-1) & \xi_1(N) & \xi_1(N-1) \end{bmatrix}$$

where $1, \ldots, N$ denotes the sampling steps. Based on (5.17) and (5.32), the model parameters $\hat{\theta}$ for offline identification can be calculated as

$$\hat{\theta} = [0.9838,\ 0.00631,\ 0.0224]^T. \tag{5.33}$$

5.3.1 Experimental Results for Battery Model Parameter On-line Identification

To show the superior performance of the proposed on-line model parameter identification algorithm based on robust RLS, the traditional RLS and offline LS algorithms are also provided here as comparisons. The initial estimated model parameter vector is randomly set as $\hat{\theta}(0) = [0.95,\ 0.001,\ 0.02]^T$ for both the robust RLS and conventional RLS. $P(0)$, N_m, and T_a are selected as $P_0 = diag\{0.1,\ 0.1,\ 0.1\}$, $N_m = 20$, and $T_a = 60$, respectively. The comparison results of the model parameters by these methods are illustrated in Fig. 5.5. The experimental voltage data of the battery (with and without outliers) and the simulated terminal voltage through the equivalent circuit model with these identified model parameters are shown in Fig. 5.6a. The corresponding simulation errors of the battery's terminal voltage (the simulated terminal voltage minus the experimental data without outliers) are illustrated in Fig. 5.6b. The average and RMS errors of these simulated terminal voltages are shown in Table 5.1, where the average and RMS errors are calculated as:

$$\begin{aligned} mean &= \tfrac{1}{N}\textstyle\sum_{k=1}^{N} |V_e(k) - V_a(k)| \\ RMS &= \sqrt{\tfrac{1}{N}\textstyle\sum_{k=1}^{N} (V_e(k) - V_a(k))^2} \end{aligned} \tag{5.34}$$

where $V_a(k)$ is the experimental data without adding outliers and $V_e(k)$ is the simulated terminal voltages. From Table 5.1, it can be seen that the equivalent circuit model with on-line identified model parameters can provide much higher accuracy than the model with offline identified model parameters. It is because that the battery model parameters vary with different operating conditions and the absence of update of the offline identified model parameters limits the model accuracy.

From Fig. 5.5, the model parameter identification results of the proposed robust RLS and the conventional RLS algorithm are very similar for the case without the insertion of outliers. But the conventional RLS algorithm has a large variation in parameter estimation caused by terminal voltage measurement outliers at 10000 s and 20000 s, while this phenomenon can be well suppressed by the proposed robust RLS algorithm. It shows that the robust RLS algorithm outperforms the conventional RLS in terms of battery model parameter identification when there exist measurement outliers, which is consistent with the above analysis.

Comparison for different initial estimated model parameters: To verify the effect of the chosen initial estimation model parameters on the parameter identification performance of the designed robust RLS method, three more experiments with different $\hat{\theta}(0)$ are constructed, which selected are $[0.9, 0.0005, 0.01]^T$, $[0.85, 0.01, 0.08]^T$, and $[0.75, 0.03, 0.12]^T$, respectively. The model parameters identified by the designed robust RLS are illustrated in Fig. 5.7 and the corresponding statistics of the terminal voltage simulation errors are shown in Table 5.2, which indicates good model parameter identification performance even in the case of large errors in the initially estimated model parameters.

Fig. 5.5 Estimation results of **a** θ_1, **b** θ_2, and **c** θ_3 using robust RLS versus RLS and offline calculation

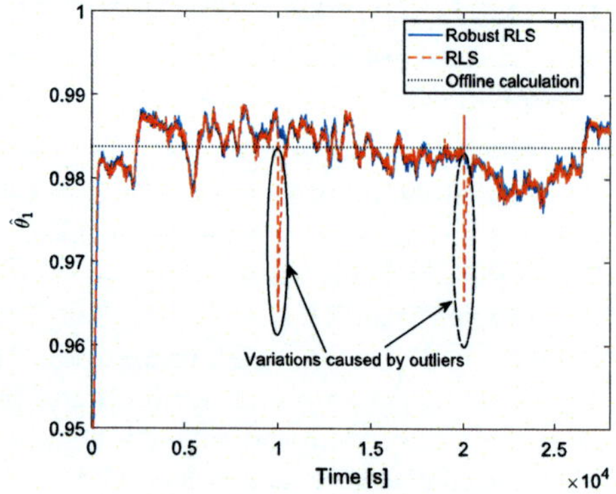

(a)

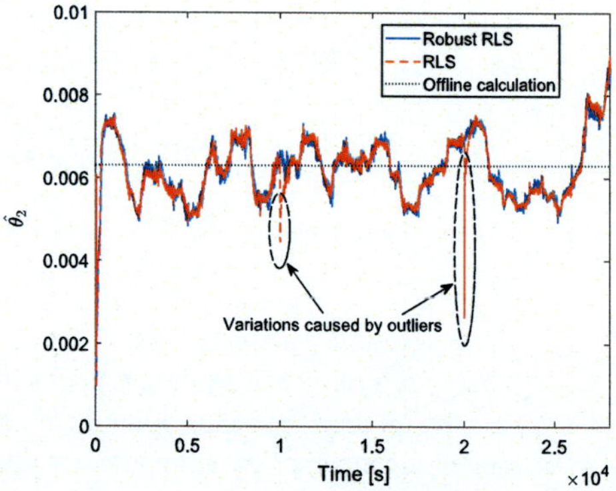

(b)

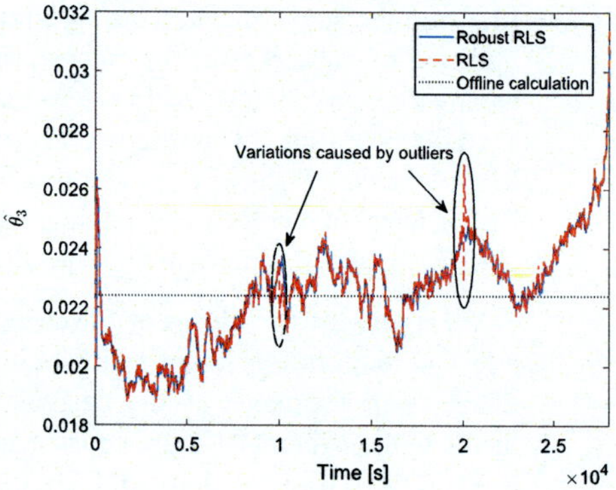

(c)

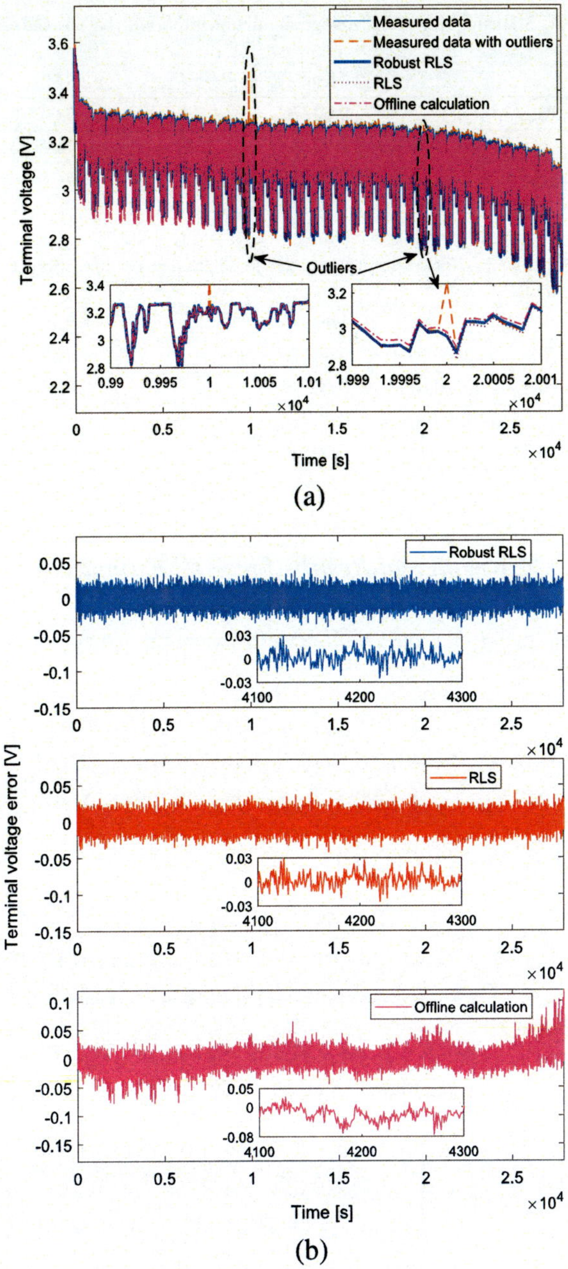

Fig. 5.6 **a** Battery's terminal voltage and simulation results of the equivalent circuit model, **b** comparison of terminal voltage simulation errors

Table 5.1 Statistics of terminal voltage simulation errors with on-line and off-line identified battery models

	Mean	RMS
Robust RLS	0.00776	0.00976
RLS	0.00781	0.00982
Offline calculation	0.0113	0.0154

Table 5.2 Comparison of terminal voltage simulation errors for different initial estimated model parameters

	Mean	RMS
$\hat{\theta}(0) = [0.95, 0.001, 0.02]^T$	0.00776	0.00976
$\hat{\theta}(0) = [0.9, 0.0005, 0.01]^T$	0.00779	0.0099
$\hat{\theta}(0) = [0.85, 0.01, 0.08]^T$	0.00792	0.0122
$\hat{\theta}(0) = [0.75, 0.03, 0.12]^T$	0.00806	0.0151

5.3.2 Experimental Results for SOC Estimation

As seen in Fig. 5.3, the relationship between the battery's OCV and SOC can be linearized as

$$V_{OC} = \begin{cases} 3.58SOC + 2.805, & \text{if } 0 \le SOC < 10\% \\ 0.213SOC + 3.142, & \text{if } 10\% \le SOC < 95\% \\ 4.682SOC - 1.103, & \text{if } 95\% \le SOC < 100\%. \end{cases} \quad (5.35)$$

According to (5.35), $C(k)$ in (5.30) can be simplified as

$$C(k) = \begin{cases} [3.58, -1] & \text{if } 0 \le S\hat{O}C(k) < 10\% \\ [0.213, -1] & \text{if } 10\% \le S\hat{O}C(k) < 95\% \\ [4.682, -1] & \text{if } 95\% \le S\hat{O}C(k) < 100\%. \end{cases} \quad (5.36)$$

With A_M selected as $diag\{1,\ 0.995\}$, the optimal gain L in the designed robust observer can be calculated in advance by solving the LMI optimization problem (5.30) as

$$L = \begin{cases} [0.2658, 0.0003] & \text{if } 0 \le S\hat{O}C(k) < 10\% \\ [3.513, 0.051] & \text{if } 10\% \le S\hat{O}C(k) < 95\% \\ [0.2072, 0.0003] & \text{if } 95\% \le S\hat{O}C(k) < 100\%. \end{cases} \quad (5.37)$$

The initial estimated state vector is randomly set as $\hat{x}(0) = [70\%,\ 0]^T$ in the proposed robust observer. The actual and estimated SOCs are shown in Fig. 5.8a, and the corresponding SOC estimation error is illustrated in Fig. 5.8b. Moreover, to

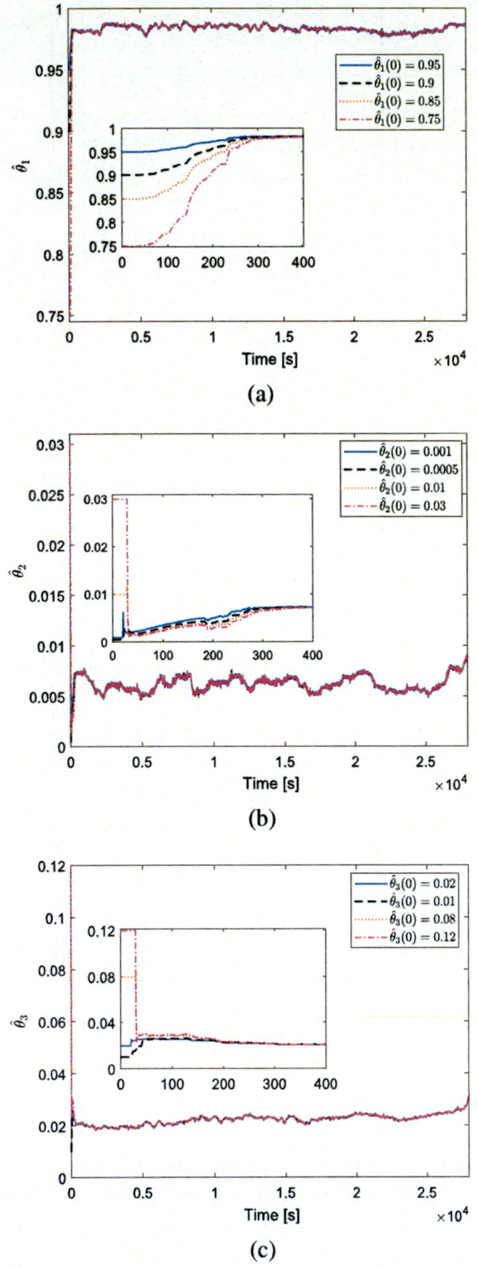

Fig. 5.7 Estimation results of **a** θ_1, **b** θ_2, and **c** θ_3 by the designed robust RLS algorithm with different initial estimated values

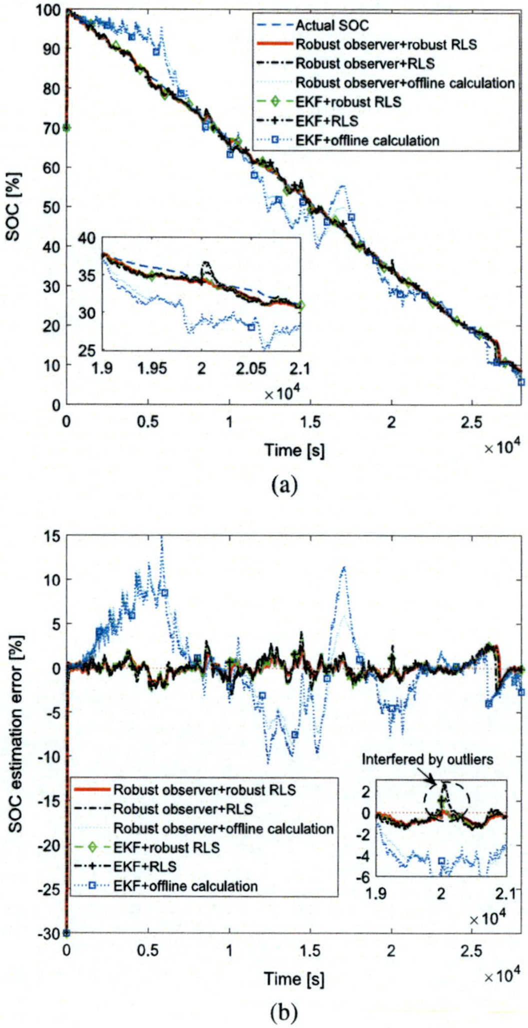

Fig. 5.8 **a** Actual and estimated SOCs, **b** SOC estimation errors

verify the superior performance of the proposed robust SOC estimation strategy, the estimation results of the EKF with robust RLS based identified model parameters, the robust observers and EKF algorithms with the conventional RLS based on-line identified model parameter and the offline calculated model parameters are also given here for comparison. From Figs. 5.5 and 5.8b, it can be observed that the SOC estimation performance of the SOC observer using the conventional RLS algorithm for estimating the model parameters is degraded because of the estimation error of the model parameters of the conventional RLS algorithm caused by the outliers of

Table 5.3 Comparisons of SOC estimation errors

	Mean (%)	RMS (%)
Robust observer + robust RLS	0.598	1.079
Robust observer + RLS	0.621	1.12
Robust observer + offline calculation	3.269	4.388
EKF + robust RLS	0.73	1.09
EKF + RLS	0.76	1.12
EKF + offline calculation	3.52	4.74

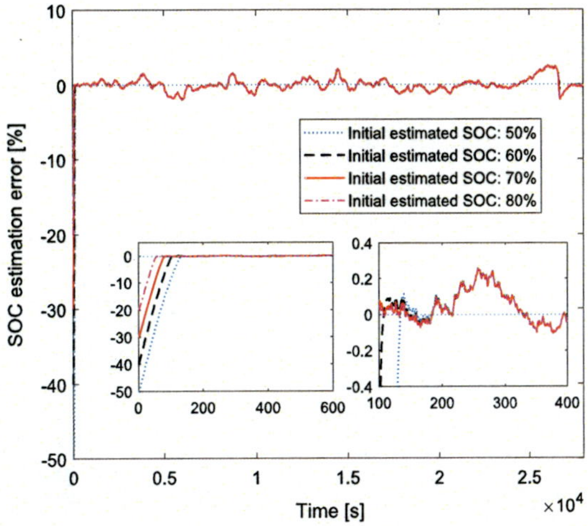

Fig. 5.9 SOC estimation errors of the proposed estimation approach with different initial estimated SOCs

the terminal voltage measurements. However, this negative impact is reduced for the proposed robust observer with on-line model parameter identified by the robust RLS, which indicates its good robustness to outliers of the battery measurement signals. As illustrated in Table 5.3, the mean and the RMS of the SOC estimation errors of the proposed robust observer with robust RLS based on-line model parameter identification are 0.598% and 1.079%, which has higher estimation accuracy compared with other algorithms.

To further validate the performance of the proposed SOC estimation strategy, the initial estimated SOC value is selected as $SOC(0) = 50\%$, $SOC(0) = 60\%$, $SOC(0) = 70\%$, and $SOC(0) = 80\%$, respectively. The corresponding SOC estimation errors are illustrated in Fig. 5.9, which shows that the proposed method can provide accurate SOC estimation even if there are unexpected outliers of the battery measurement signal in practice, which verifies its great robustness and superior SOC estimation performance.

Current including low-current excitation: Finally, a current profile including a rest for 3000 s and a constant current discharging for 2000 s is utilized to verify the

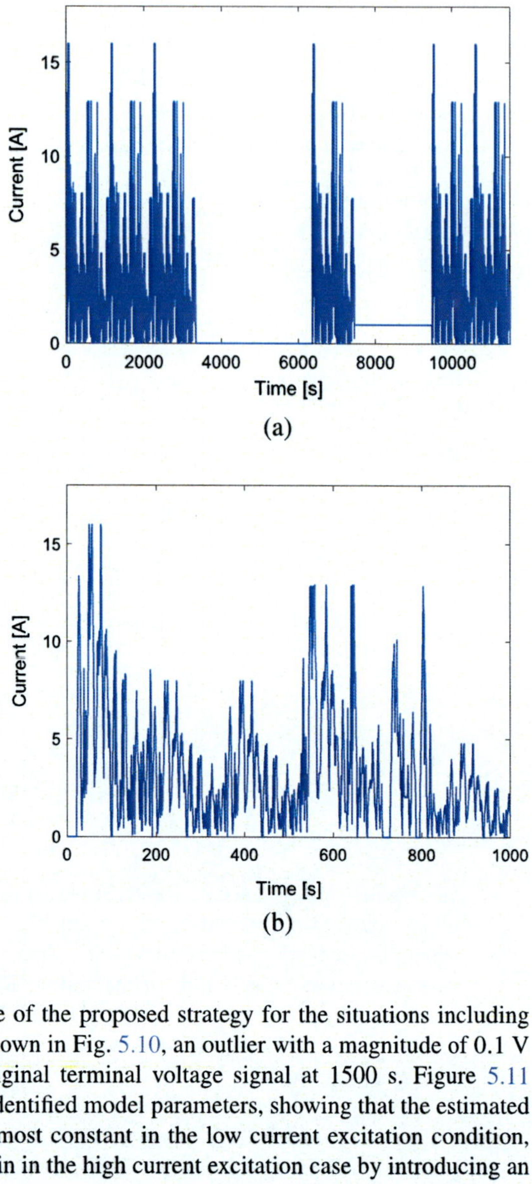

Fig. 5.10 **a** Current profile including low-current excitation, **b** zoom of the applied current

SOC estimation performance of the proposed strategy for the situations including low-current excitation. As shown in Fig. 5.10, an outlier with a magnitude of 0.1 V is added to the battery's original terminal voltage signal at 1500 s. Figure 5.11 illustrates the results of the identified model parameters, showing that the estimated model parameters remain almost constant in the low current excitation condition, while convergence starts again in the high current excitation case by introducing an additional loop as in (5.31).

The SOC estimation errors are depicted in Fig. 5.12 and Table 5.4, showing that the SOC estimation error increases slowly for the discharging condition with a constant current. This is because changes in model parameters cannot be tracked during this period. However, it starts to converge again once the current excitation recovers.

Fig. 5.11 Estimation results of **a** θ_1, **b** θ_2, and **c** θ_3 using robust RLS versus RLS and offline calculation for current profile including low-current excitation

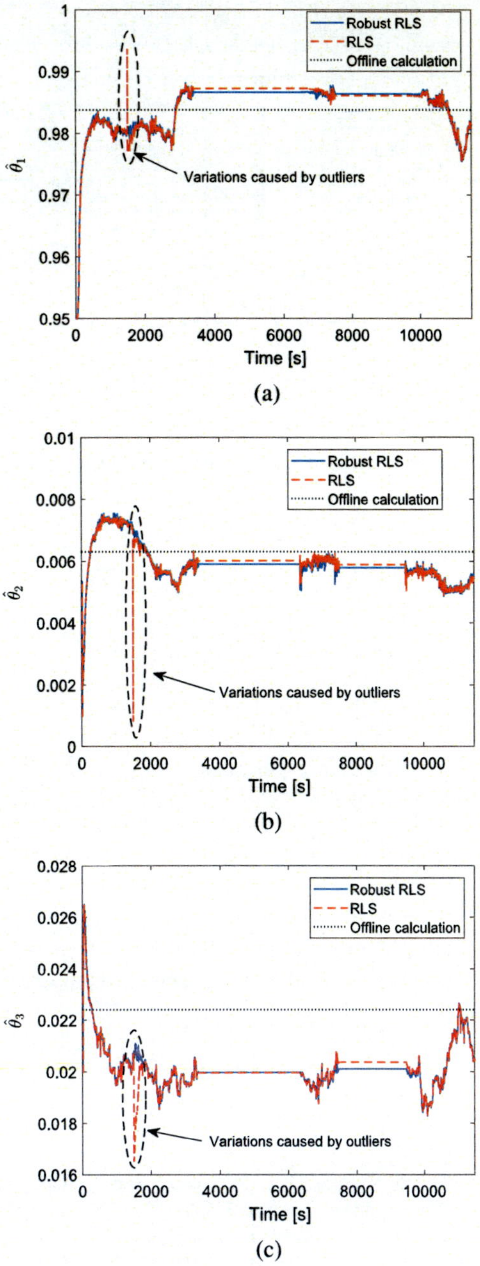

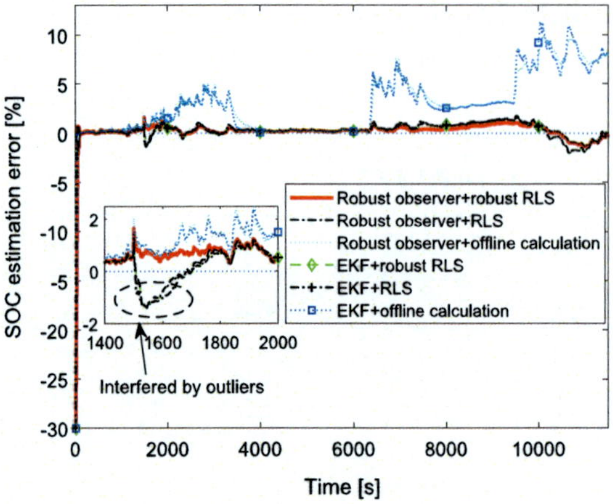

Fig. 5.12 SOC estimation errors for current including low-current excitation

Table 5.4 Comparisons of SOC estimation errors

	Mean (%)	RMS (%)
Robust observer + robust RLS	0.489	1.302
Robust observer + RLS	0.59	1.361
Robust observer + offline calculation	2.981	4.187
EKF + robust RLS	0.55	1.8
EKF + RLS	0.6	1.83
EKF + offline calculation	3.03	4.15

The experimental results show that the proposed method can accurately estimate the battery's SOC even for the current including low-current excitation. In addition, it can also be concluded that the proposed strategy outperforms the other five SOC estimation algorithms from the comparison results in Fig. 5.12 and Table 5.4.

References

1. Q. Ouyang, J. Chen, and J. Zheng, "State of charge observer design for batteries with on-line model parameter identification: A robust approach," *IEEE Transactions on Power Electronics*, vol. 35, no. 6, pp. 5820–5831, 2020.
2. A. Vahidi, A. Stefanopoulou, and H. Peng, "Recursive least squares with forgetting for online estimation of vehicle mass and road grade: theory and experiments," *International Journal of Vehicle Mechanics and Mobility*, vol. 43, no. 1, pp. 31–35, 2005.

3. N. Zhou, J. W. Pierre, D. J. Trudnowski, and R. T. Guttromson, "Robust RLS methods for online estimation of power system electromechanical modes," *IEEE Transactions on Power Systems*, vol. 22, no. 3, pp. 1240–1249, 2007.
4. B. Kovaevi, M. Milosavljevi, and M.Veinovi, "Robust recursive AR speech analysis," *Signal Processing*, vol. 44, no. 2, pp. 125–138, 1995.
5. R. M. Johnstone, C. R. Johnson, R. R. Bitmead, and B. D. O. Anderson, "Exponential convergence of recursive least squares with exponential forgetting factor," in *IEEE Conference on Decision and Control*, 1982, pp. 994–997.
6. H. K. Khalil, *Nonlinear Systems*, 3rd ed. Englewood Cliffs, NJ, USA: Prentice-Hall, 2002.
7. F. Zhang, *The Schur Complement and Its Applications*. New York, NY, USA: Springer, 2005.
8. D. Simon, *Optimal State Estimation: Kalman,* H_∞*, and Nonlinear Approaches*. Hoboken, NJ, USA: Wiley, 2006.
9. L. Xie, C. E. D. Souza, and Y. Wang, "Robust filtering for a class of discrete-time uncertain nonlinear systems: An H_∞ approach," *International Journal of Robust and Nonlinear Control*, vol. 6, no. 4, pp. 297–312, 1996.
10. K. Zhou and J. Doyle, *Essentials of Robust Control*. Englewood Cliffs, NJ, USA: Prentice Hall, 1998.

Chapter 6
User-Involved Battery Charging Control with Economic Cost Optimization

Most of the existing charging control methods focus on fast charging to accelerate the charging speed with guaranteeing the batterys safety. When the charging demand is urgent, a large current is inevitable to be used to achieve fast charging. However, it is not uncommon to encounter scenarios where charging times are sufficient in practice, such as batteries being charged overnight at home. For these cases, it is better to provide a low but healthy charging current for the battery rather than a high but harmful charging current. Considering the user's charging demand in the charging control method, the charging current can be self-adjusted according to user specifications and battery dynamics, which not only makes the charger smarter, but also reduces the capacity loss of the battery. Based on this idea, it is valuable to consider the user's charging demand in the battery charging method.

In this chapter, based on the coupled electro-thermal model, a multi-objective constrained optimization-based charging control method [1] is proposed that considers user demand realization, economic cost optimization, and energy loss reduction in the charging process. It can bring the benefits of more user-friendly service, less electricity bills, and less energy waste.

6.1 Battery Model and Constraints

6.1.1 Battery Model

For model-based charging control design, a coupled electro-thermal model [2] is used to characterize the dynamics of the cylindrical lithium-ion battery. As illustrated in Fig. 6.1, the voltage source V_{OC} and the resistance R_0 are utilized to simulate the battery's energy storage and charge/discharge energy loss, respectively. The RC networks (R_1, C_1) and (R_2, C_2) are used to characterize the battery's short-term and

Q. Ouyang and J. Chen, *Advanced Model-Based Charging Control for Lithium-Ion Batteries*, https://doi.org/10.1007/978-981-19-7059-7_6

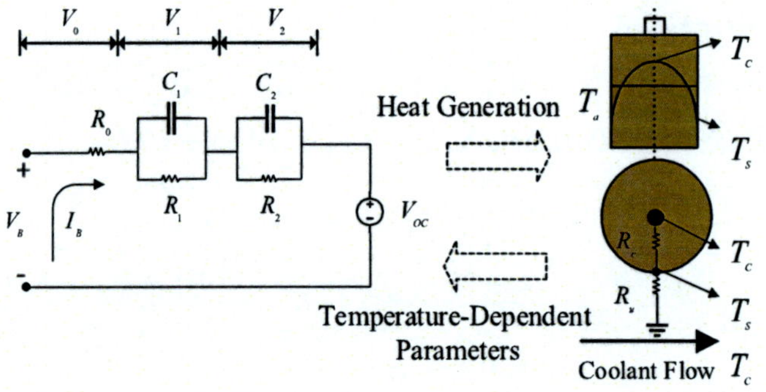

Fig. 6.1 Coupled electro-thermal model of the cylindrical lithium-ion battery [2]

long-term transient responses. According to Kirchhoff's laws of current and voltage, the electrodynamics of the battery can be described as

$$\begin{aligned} \dot{SOC} &= \frac{1}{3600C_n} I_B \\ \dot{V}_1 &= -\frac{1}{R_1C_1} V_1 + \frac{1}{C_1} I_B \\ \dot{V}_2 &= -\frac{1}{R_2C_2} V_2 + \frac{1}{C_2} I_B \\ V_B &= V_{OC} + V_1 + V_2 + R_0 I_B \end{aligned} \tag{6.1}$$

where SOC, C_n, I_B, and V_B are the SOC, nominal capacity, charging current, and terminal voltage of the battery, respectively. The open circuit voltage of the battery V_{OC} is a nonlinear function of SOC and can be expressed as $V_{OC} = g(SOC)$. V_1 and V_2 represent the voltages across the capacitors C_1 and C_2, respectively . T_c and T_s represent the core and surface temperatures of the battery, which can be calculated according to the principle of energy conservation as [3]

$$\begin{aligned} \dot{T}_c &= \frac{T_s - T_c}{R_c C_c} + \frac{Q}{C_c} \\ \dot{T}_s &= \frac{T_{amb} - T_s}{R_u C_s} - \frac{T_s - T_c}{R_c C_s} \end{aligned} \tag{6.2}$$

where T_{amb} denotes the ambient temperature of the battery; R_c and R_u represent the heat conduction and convection resistance; C_c and C_s represent the battery's internal and surface heat capacities, respectively; Q represents the heat generated and is calculated as

$$Q = I_B(V_1 + V_2 + R_0 I_B). \tag{6.3}$$

The battery's temperature T_a is defined as the average of T_s and T_c which can be denoted as

$$T_a = \frac{1}{2}(T_s + T_c). \tag{6.4}$$

Note that the electrical and thermal dynamics of the battery in (6.1)–(6.2) are strongly coupled, where the resistances and capacitances R_0, R_1, R_2, C_1, and C_2 in (6.1) are nonlinear functions of the battery's SOC and temperature, while the heat generation Q in (6.2) is influenced by the electrodynamics of the battery. To facilitate the implementation of the charging control algorithm in digitally controlled systems, the battery model (6.1)–(6.2) can be discretized using the Euler method, which keeps the charging current constant during each sampling interval T. To simplify the notation, the state vector $x(k) = [x_1(k), x_2(k), \ldots, x_5(k)]^T \in \mathbb{R}^5$, the output vector $y(k) = [y_1(k), y_2(k), y_3(k)]^T \in \mathbb{R}^3$, and the input $u(k) \in \mathbb{R}$ are defined as

$$\begin{aligned} x(k) &\triangleq [SOC(k), V_1(k), V_2(k), T_c(k), T_s(k)]^T \\ y(k) &\triangleq [SOC(k), V_B(k), T_a(k)]^T \\ u(k) &\triangleq [I_B(k)]. \end{aligned} \tag{6.5}$$

The battery's coupled electro-thermal model can be expressed as the following discrete-time state-space formulation

$$\begin{aligned} x(k+1) &= f(x(k), u(k)) \\ y(k) &= h(x(k), u(k)) \end{aligned} \tag{6.6}$$

where $f(\cdot) \in \mathbb{R}^5$ and $h(\cdot) \in \mathbb{R}^3$, respectively, are

$$\begin{aligned} f(\cdot) &= \begin{bmatrix} x_1(k) + \frac{T}{3600C_n}u(k) \\ (1 - \frac{T}{R_1C_1})x_2(k) + \frac{T}{C_1}u(k) \\ (1 - \frac{T}{R_2C_2})x_3(k) + \frac{T}{C_2}u(k) \\ (1 - \frac{T}{R_cC_c})x_4(k) + \frac{Tx_5(k)}{R_cC_c} + \frac{T(x_2(k)+x_3(k)+R_0u(k))u(k)}{C_c} \\ \frac{T}{R_cC_s}x_4(k) + (1 - \frac{T}{R_cC_s} - \frac{T}{R_uC_s})x_5(k) + \frac{TT_{amb}}{R_uC_s} \end{bmatrix} \\ h(\cdot) &= \begin{bmatrix} x_1(k) \\ g(x_1(k)) + x_2(k) + x_3(k) + R_0u(k) \\ \frac{1}{2}(x_4(k) + x_5(k)) \end{bmatrix}. \end{aligned} \tag{6.7}$$

6.1.2 *Safety-Related Constraints*

Since a large charging current can cause irreversible damage to the battery, it should be kept within the suitable range recommended by the battery's instruction manual. It can be expressed as:

$$0 \leq u(k) \leq u_M \tag{6.8}$$

where $u_M \in \mathbb{R}$ is the maximum charging current allowed for the battery. In addition, overcharging and overheating can cause an accelerated decrease in battery's capacity and even lead to safety issues. To avoid these issues, the battery's SOC, terminal voltage, and temperature must be prevented from exceeding their allowable limits. It generates

$$y(k) \leq y_M \tag{6.9}$$

where $y_M \in \mathbb{R}^3$ denotes the upper limit of the battery output vector.

6.2 Charging Tasks

The cost functions for multiple charging tasks, including user-involved task, economic cost optimization, and energy loss reduction, are developed here.

6.2.1 *User-Involved Charging Task*

In the user-involved charging method, depending on the future demand, users can pre-set their target SOC SOC_d and charging duration T_d of the battery, respectively. After charging, the SOC of the battery can be maximally close to SOC_d. In view of this, the cost function J_1 with respect to the user demand can be expressed as

$$J_1 = (x_1(N) - SOC_d)^2 \text{ with } T_d = NT \tag{6.10}$$

where N is the number of sampling steps. To keep the computational burden consistent for users' different specified charging duration T_d, N is fixed in the charging control algorithm and the sampling period can be computed as $T = \frac{T_d}{N}$. Note that user requirements may not always be achievable in practice because safety-related constraints (6.8) and (6.9) must be satisfied during the charging process. Hence, we try to drive $x_1(N)$ toward SOC_d to the greatest extent through minimizing the cost function J_1 in (6.10).

6.2.2 *Economic Cost Optimization*

The economic cost optimization is another necessary and crucial objective of battery charging in practical applications. The main economic cost is the money spent on electricity. The electrical price around the world usually fluctuates with the peak or valley of electricity consumption during a day, defined as the peak-valley time-of-use (TOU) price [4]. Since the peak and valley electricity prices vary widely, it is valuable and necessary to adjust the charging pattern within the pre-set charging period to reduce the economic cost of the consumed electricity. The consumed electricity fee can be computed as

$$J_2 = \sum_{k=0}^{N-1} p(k) \frac{u(k)y_2(k)T}{3.6 \times 10^6} \tag{6.11}$$

where $p(k)$ is the TOU electricity price in kWh at the sampling step k.

6.2.3 Energy Loss Reduction

As part of the electrical energy is dissipated as waste heat rather than being converted into chemical energy stored in the battery, it is of great important to improve the charging efficiency by reducing the amount of energy loss of the battery during the charging process. Based on (6.3), the cost function on energy loss, denoted as J_3, can be calculated as

$$J_3 = \sum_{k=0}^{N-1} u(k)(x_2(k) + x_3(k) + R_0(k)u(k))T. \tag{6.12}$$

Note that J_3 is also related to the battery's temperature rise. By constraining it, the temperature rise of the battery can be suppressed at the same time.

6.2.4 Multi-objective Formulation

In the designed battery charging control strategy, user demand realization, economic cost optimization, and energy loss reduction are considered. To strike a balance between these three objectives, a compromise solution can be obtained by using a weighted metric method [5]. Due to different units and scales, the cost functions J_1, J_2, and J_3 need to be normalized to a range between 0 and 1 utilizing the min-max normalization algorithm as

$$J_i' = \frac{J_i - J_{i_m}}{J_{i_M} - J_{i_m}} \tag{6.13}$$

where J_i', J_{i_m}, and J_{i_M} $(1 \leq i \leq 3)$ are the normalized, minimum, and maximum values of J_i, respectively. From (6.10)–(6.13), the charging control tasks can be formulated to minimize the following cost functions:

$$J = \gamma_1 J_1' + \gamma_2 J_2' + \gamma_3 J_3' \tag{6.14}$$

where $\gamma_1 > 0$, $\gamma_2 > 0$, and $\gamma_3 > 0$ are weight factors whose values indicate the relative importance of the corresponding charging targets. Note that most of the existing charging strategies focus on fast charging to fully charge the battery in the shortest possible time. But these results are not optimal for cases when the charging time required by the user is sufficient. To remedy this, the multiple objectives of user demand realization, economic cost optimization, and energy loss reduction are considered in the charging control algorithm as (6.14), and the charging current can be automatically adjusted with the user demand and TOU electricity price, which can satisfy the user's charging demand while reduce the users' economic burden.

6.3 Optimal Battery Charging Control Design

The schematic diagram of the designed multi-objective optimal charging control method is shown in Fig. 6.2. First, the charging device receives the user's demand and considers demand, economic cost optimization, energy loss reduction, and safety-related constraints. Next, a constrained optimization issue is formulated and solved using the barrier method to obtain the optimal charging current. Finally, the charger provides the corresponding current to the battery.

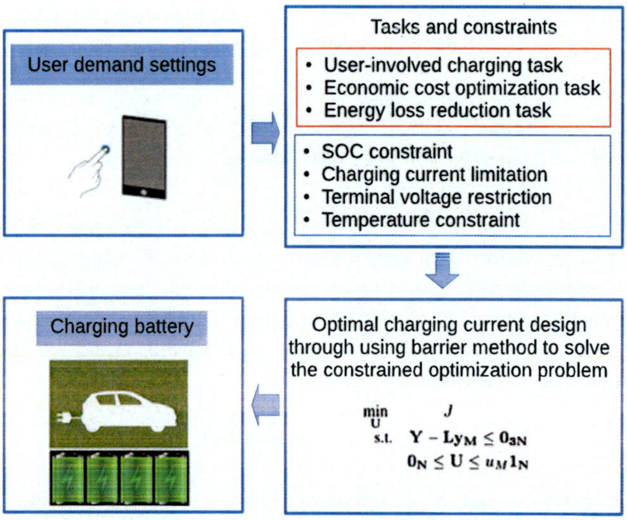

Fig. 6.2 Framework for the multi-objective optimal battery charging control method

6.3.1 Optimal Charging Control Algorithm

Considering the multi-objective cost function for user-involvement, economic cost optimization, energy loss reduction in (6.14), and the safety-related constraints in (6.8) and (6.9), from the battery model (6.6), the battery charging control algorithm can be developed as follows:

$$\begin{aligned}\min_{u(0),u(1),\ldots,u(N-1)} \quad & J \\ \text{s.t.} \quad & x(k+1) = f(x(k), u(k)), \;\; x(0) \\ & h(x(k+1), u(k+1)) \leq y_M \\ & 0 \leq u(k) \leq u_M\end{aligned} \tag{6.15}$$

where $x(0)$ denotes the initial state vector of the battery. With defining the input sequence $U \triangleq [u(0), u(1), \ldots, u(N-1)]^T \in \mathbb{R}^N$, by iteration, $x(k)$ and $y(k)$ $(1 \leq k \leq N)$ can be rewritten as

$$\begin{aligned}x(k) &\triangleq s_k(x(0), U) \\ &= f(f(\ldots f(x(0), u(0)), \ldots u(k-2)), u(k-1)) \\ y(k) &\triangleq m_k(x(0), U) \\ &= h(s_k(x(0), U), u(k)).\end{aligned} \tag{6.16}$$

Next, the output sequence $Y \triangleq [y^T(1) \ldots, y^T(N)]^T \in \mathbb{R}^{3N}$ can be expressed as

$$Y = \begin{bmatrix} m_1(x(0), U) \\ \cdots \\ m_N(x(0), U) \end{bmatrix}. \tag{6.17}$$

Based upon (6.17), (6.15) can be represented as

$$\begin{aligned}\min_{U} \quad & J \\ \text{s.t.} \quad & Y - Ly_M \leq 0_{3N} \\ & 0_N \leq U \leq u_M 1_N\end{aligned} \tag{6.18}$$

with $L = [I_3, \ldots, I_3]^T$, where 1_N and 0_N are column vectors with N ones and zeros, respectively; I_3 denotes a identity matrix with dimension 3×3.

6.3.2 *Optimal Charging Current Determined by Barrier Method*

Since (6.18) is a standard constrained nonlinear optimization issue, the powerful barrier method [6] can solve it by approximately formulating it as an unconstrained minimization issue to which Newton's method can be applied. By utilizing the convex barrier function [6] to replace the inequality constraint, (6.18) can be transformed into the following unconstrained optimization formulation:

$$\min_{U} \quad J - \frac{1}{\mu}\sum_{j=1}^{5N} \log(-G_j) \tag{6.19}$$

where μ is a positive parameter, $-\sum_{j=1}^{5N} \log(-G_j)$ denotes the logarithmic barrier function, and G_j is the j-th element of the vector $G \in \mathbb{R}^{5N}$ that

$$G = \begin{bmatrix} Y - Ly_M \\ U - u_M 1_N \\ -U \end{bmatrix}.$$

From (6.19), it is observed that the logarithmic barrier function increases without bound if any $G_j \to 0$ $(1 \leq j \leq 5N)$. It can guarantee that the optimal solution is within the feasible range of $\{U \in \mathbb{R}^N | G \leq 0_{5N}\}$. It should be pointed out that (6.19) is an approximation of the original problem (6.18), where the approximation accuracy increases with the parameter μ. But (6.19) is difficult to be solved by Newton's method if μ is chosen too large, because it causes its Hessian matrix to change rapidly near the boundary of the feasible range.

To resolve this conflict, this problem is transferred to solving a series of issues of the form (6.19) by Newton's method with increasing μ at each step, and the solution for this problem with the previous value of μ is treated as the initial point for the next iteration [6]. The detailed procedure of the barrier method is illustrated in Algorithm 6.1. By employing the barrier method, the sequence of the optimal charging currents $u(k)$ $(0 \leq k \leq N-1)$ can be off-line calculated.

Algorithm 6.1: Barrier method [6]

---

1. Set the initial optimization variable $U = U^{(0)}$, parameters $\mu > 0$, $c > 1$, and tolerance $\varepsilon > 0$.
2. Use the Newton's method to solve (6.19) with the initial value U to obtain the optimal U^*.
3. Update $U = U^*$.
4. Stop and output U, if $\frac{5N}{\mu} < \varepsilon$. Otherwise, set $\mu = c\mu$ and return to Step (2).

Note that the barrier method is a gradient-based strategy that can employ the gradient of the function to be optimized to calculate the search direction of its solution.

Compared with the commonly utilized heuristic optimization algorithm such as the genetic algorithm (GA) and the particle swarm optimization (PSO) algorithm, it needs less computation amount and has a faster convergence speed, which is more suitable for battery charging control.

6.4 Simulation Results

To validate the performance of the designed user-involved battery charging control method with economic cost optimization, extensive simulations were conducted in MATLAB 2018b with a 3.6 GHz Intel i3-8100 CPU. The nominal capacity of the battery is 2.3 Ah, and the ambient temperature is 25 °C. The battery coupled electro-thermal model parameters are seen in [7] and are not repeated here for brevity. The upper limitations of the charging current, SOC, terminal voltage, and temperature of the battery are chosen as 6.9 A (3-C rate), 100%, 3.6 V, and 45 °C, respectively. The TOU electricity price in Beijing from [8] is utilized here as shown in Fig. 6.3. The peak electricity price is 1.253 Yuan/kWh within 10:00–15:00 and 18:00–21:00. The flat electricity price is 0.781 Yuan/kWh within 07:00–10:00, 15:00–18:00, and 21:00–23:00. The valley electricity price is 0.335 Yuan/kWh within 23:00–07:00.

In the proposed battery charging control algorithm, the lower bounds of the cost functions J_{1_m}, J_{2_m} and J_{3_m} are all set to zero. J_{1_M} is chosen as 1 because the battery's SOC is within the range of [0, 100%]. J_{2_M} is chosen as 0.0104 Yuan, which is calculated as 2.3 Ah × 3.6 V × 0.001 × 1.253 Yuan/kWh with the maximum electricity price of 1.253 Yuan/kWh. J_{3_M} is 1438.4 J, which is calculated by charging the battery from 0 SOC to 100% energy loss in CC-CV mode with a current of 3-C rate. The weight coefficients γ_1, γ_2, and γ_3 are selected as 1, 0.01, and 0.001, respectively. The number of sampling steps is chosen as $N = 10$.

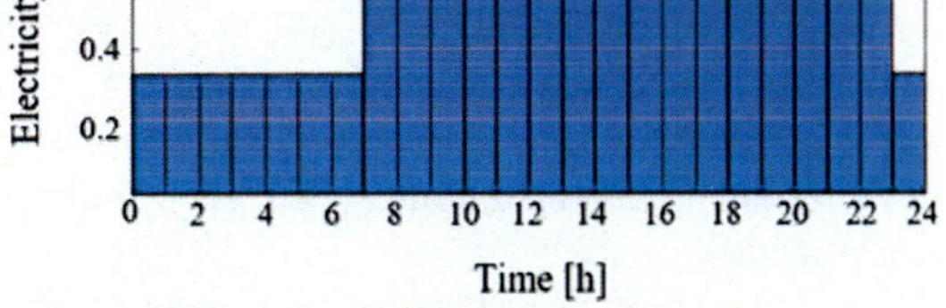

Fig. 6.3 The TOU electricity price in Beijing [8]

6.4.1 Charging Results

The target SOC of the user is set to 100%, and the desired charging duration is selected from 6:00 to 8:00 as a case study. The battery's initial SOC is 10%. The simulation results for the battery's SOC, current, terminal voltage, and temperature are shown in Fig. 6.4a–d, respectively. It is observed that the SOC of the battery can be charged to 99.72% with all charging constraints guaranteed at the end of the charging process. The electricity fee and energy loss are 0.00237 Yuan and 544.97 J, respectively. As illustrated in Fig. 6.4b, the charging current can be self-adjusted with the fluctuation of electricity price, where a large charging current is used during the first hour (6:00–7:00 with a low electricity price 0.335 Yuan/kWh), while the charging current is almost zero during 7:00–8:00 with a high electricity price of 0.781 Yuan/kWh. This operation can effectively reduce the electricity cost of battery charging.

6.4.2 Comparison with Other Commonly Used Optimization Algorithms

Two heuristic optimization algorithms, GA and PSO, are also used here as comparisons. The size of population/particles, the length/dimension of a chromosome/particle, and the maximum number of generations/iterations are chosen as 30, 10, and 600 in the GA/PSO, respectively. Table 6.1 shows the charging results, indicating that they have similar performance with respect to the gap between the actual and desired SOCs, electricity cost, and energy loss at the end of the charging process. However, the barrier method used requires much less computing time than GA and PSO, which is more suitable for charging control of the battery.

6.4.3 Comparison with Charging Control Strategy without Economic Cost Optimization

In order to compare the performance of the charging method without economic cost optimization, γ_2 is set to zero in our designed charging method and the corresponding charging results are provided as in Fig. 6.4. After 2 h of charging, the battery SOC can reach 99.85% and the energy loss is 314.97 J. As illustrated in Fig. 6.4b, since the charging current cannot vary with the TOU electricity price to reduce the economic cost, the consumed electricity cost is 0.0038 Yuan of the charging method without economic cost optimization, which is 1.58 times that of our proposed charging control method.

Taking an EV battery pack consisting of 7000 batteries as an example (e.g., Tesla), assuming that it is charged 30 times per year in this mode, the electricity fees with

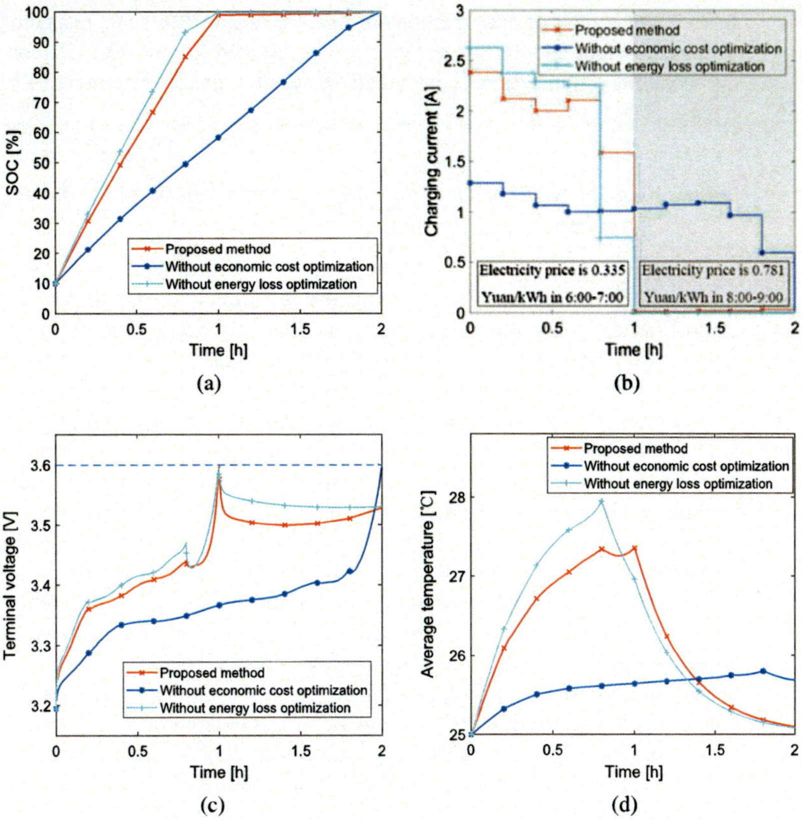

Fig. 6.4 Comparison of **a** SOC, **b** charging current, **c** terminal voltage, and **d** temperature for the methods with and without economic cost/energy loss optimization

Table 6.1 Simulation results for different optimization algorithms

	SOC gap (%)	Electricity fee (Yuan)	Energy loss (J)	Computing time (s)
Barrier method	0.28	0.00237	544.97	41.07
GA	0.15	0.00241	556.37	443.33
PSO	0.22	0.00236	561.56	339.49

and without economic optimization are 497.7 Yuan and 798 Yuan, respectively. The result shows that 300.3 Yuan per year can be reduced for one EV with considering the economic optimization. It can greatly reduce the financial burden of EV users.

6.4.4 Comparison with Charging Control Strategy Without Energy Loss Optimization

In order to show the energy waste reduction performance of the designed charging control strategy, the battery charging method without energy loss optimization (by setting $\gamma_3 = 0$ in our designed charging method) is provided as a comparison. The comparison results are shown in Fig. 6.4, where the energy losses under our designed battery charging control method and the strategy without energy loss optimization are 544.97 J and 587.07 J, respectively. The results denote that the energy loss can be reduced by 7.2% by considering the its reduction task in our charging control method. In addition, it can be seen from Fig. 6.4d that the temperature rise of 0.58 °C can be reduced by utilizing the proposed charging control method with energy loss optimization.

We also provide the comparison results of the energy loss at different charging rates under the strategies with and without the energy loss optimization, where the charging process starts at 7:00 and the charging demand settings is 1 h, 1.5 h , and 2 h, respectively. As shown in Fig. 6.5, the energy loss J_3 under our designed charging control method is 561.9 J, 398.5 J, and 310.3 J, respectively. It is observed that the energy loss increases when the charging time decreases. Compared with the method without energy loss optimization, the energy loss can be reduced by 10.8 J, 4.5 J and 2.7 J, respectively.

6.4.5 Simulation Results for Different Weight Selections

In the designed battery charging method, the magnitudes of the weight coefficients γ_1, γ_2 and γ_3 in (6.14) show the relative importance of user demand satisfaction, economic cost optimization and energy derogation tasks, respectively. Since satisfying the user's charging demand is the most important goal, γ_1 is chosen much larger than the other two weights. $\gamma_1 = 1$, and γ_2 and γ_3 are, respectively, chosen to be 0, 0.001, 0.005, 0.01, and 0.05 in the simulation to verify the weight selection on the charging performance. The corresponding difference between the actual SOC and the desired SOC, the consumed electricity cost, the energy loss, and the charging current are shown in Fig. 6.6. It shows that a larger γ_2 can bring less electricity cost but higher energy loss, which is inversely proportional to the value of γ_3. From the results, the choice of $\gamma_2 = 0.01$, $\gamma_3 = 0.001$ shows a proper balance between these conflicting objectives. Therefore, they are selected in our designed charging control strategy.

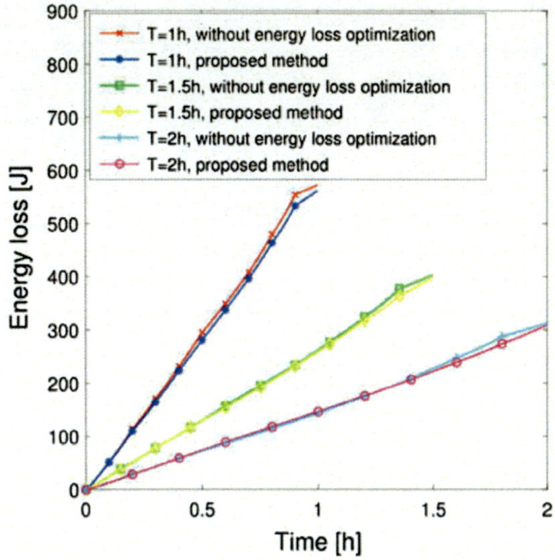

Fig. 6.5 Comparison of energy loss at different charging speeds

6.4.6 Simulation Results for Different User Demands

To further verify that the proposed charging method can be applied to satisfy the needs of different users, we randomly selected the required charging duration to perform more tests, as shown in cases 1–5 in Table 6.2 and Fig. 6.7, where the charging duration varies from 0.5 to 8 h. Figures 6.8 and 6.9 show the simulation results of the SOC response and charging current for all cases, respectively. The difference between the actual and desired SOC, the electricity cost, the energy loss at the end of the charging process, and the average charging current are shown in Table 6.3. It denotes that a large/small charging current is designed to achieve a tight/sufficient preset charging duration with the user requirements considered in the proposed charging strategy, which is consistent with the above analysis. Moreover, most of the charging operations are concentrated in the interval of low TOU electricity price during the preset charging period, while the charging current is almost zero during the period with high electricity price. The electricity cost can be effectively reduced, reflecting the superior performance of the multi-objective charging strategy designed with considering economic cost optimization.

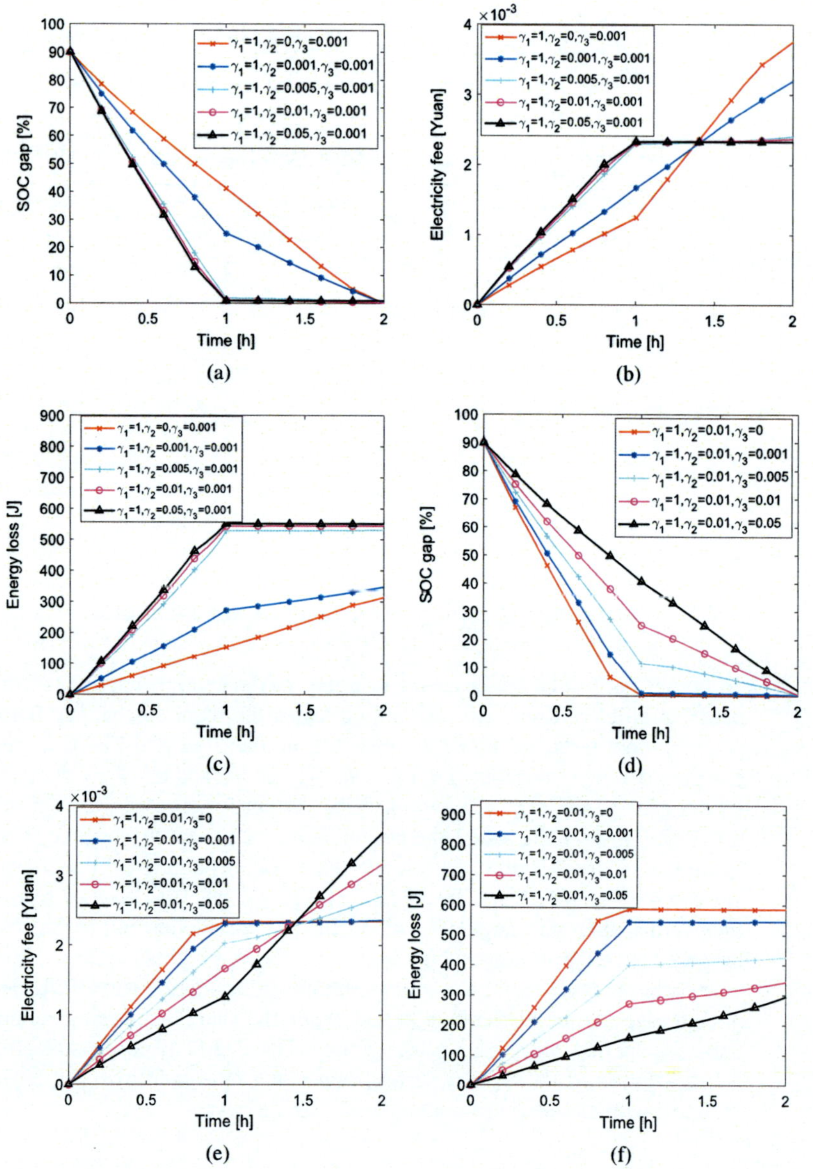

Fig. 6.6 Responses of **a** SOC gap, **b** electricity fee, and **c** energy loss for different γ_2; **d** SOC gap, **e** electricity fee, and **f** energy loss for different γ_3

Table 6.2 Different user demands

Case 1	Case 2	Case 3	Case 4	Case 5
16:00–16:30	12:00–13:00	9:00–13:00	13:00–19:00	20:00–4:00 (next day)

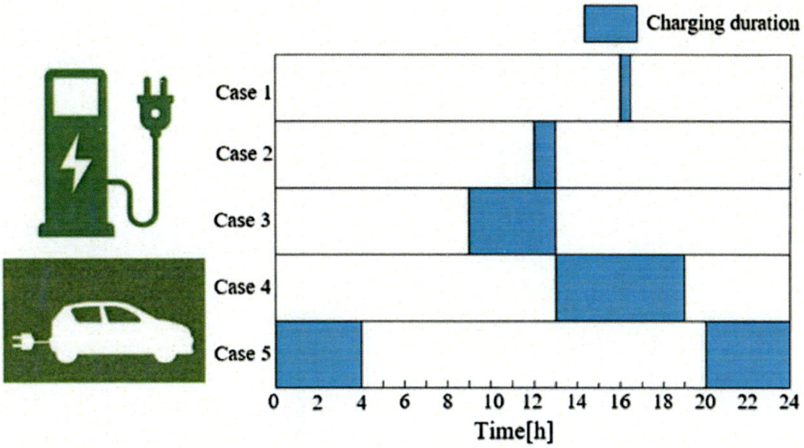

Fig. 6.7 Illustration of different user demands

6.4.7 Comparison with Traditional CC-CV Charging Methods

The charging results of the conventional CC-CV method are also provided for comparison, where its constant charging current is set to 1 C-rate, 2 C-rate and 3 C-rate, respectively. The cases with the shortest and longest charging times, i.e., Cases 1 and 5, are chosen as examples. The corresponding SOC gap between the actual SOC and the desired SOC, electricity cost, and energy loss during charging are illustrated in Table 6.4. For the demand in Case 1, the SOC and charging current under the above-mentioned CC-CV method are shown in Fig. 6.10. For the CC-CV method with 1 C-rate charging current, it brings less electricity fee and energy loss compared to charging currents of 2 C-rate and 3 C-rate for Case 5, but it cannot fully charge the battery for Case 1. With 3 C-rate charging current, although the battery can be fully charged in all cases, it incurs more than 3.9 times electricity costs and more than 9.5 times energy losses compared to our designed charging control strategy for Case 5. These defects are due to the fact that the charging current cannot be self-adjusted along with the user demand and TOU electricity price in the traditional CC-CV strategy. It demonstrates the superior performance of our designed charging strategy by comprehensively considering the multiple objectives of user demand, economic energy optimization, and energy loss reduction.

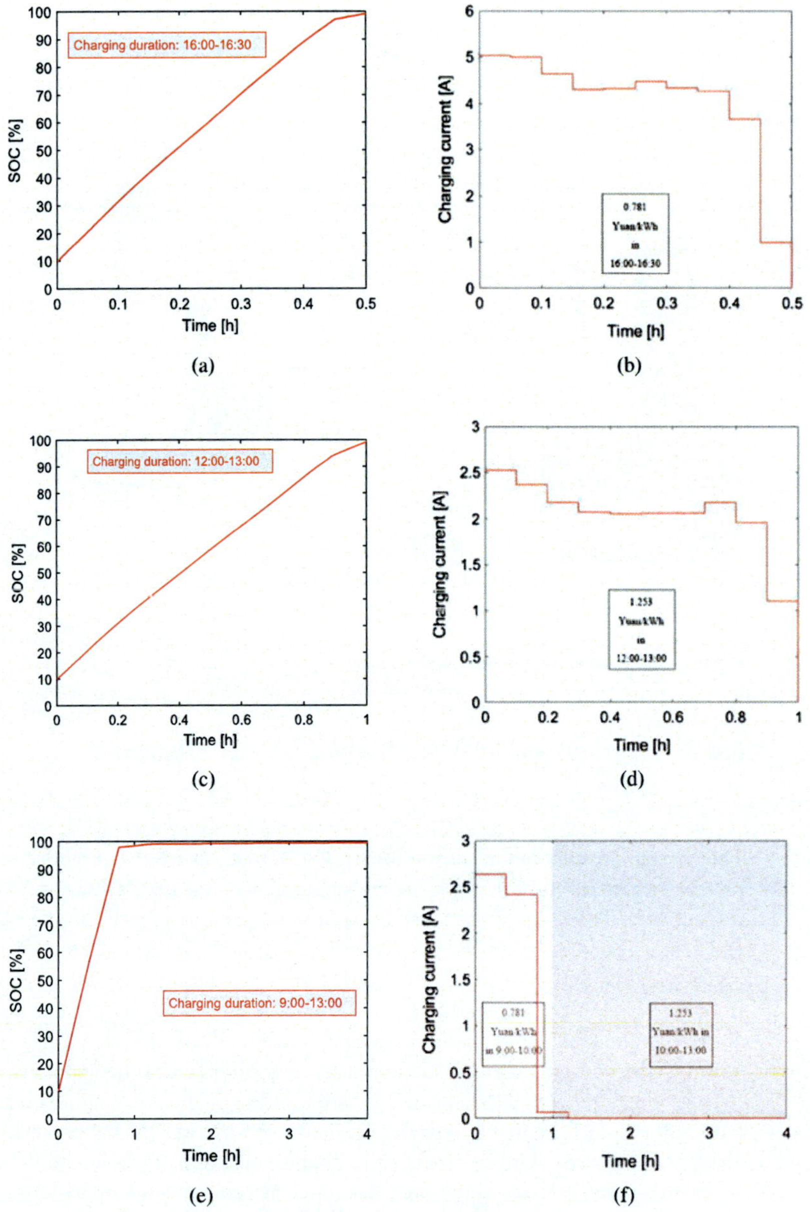

Fig. 6.8 **a** SOC response and **b** charging current for Case 1; **c** SOC response and **d** charging current for Case 2; **e** SOC response and **f** charging current for Case 3

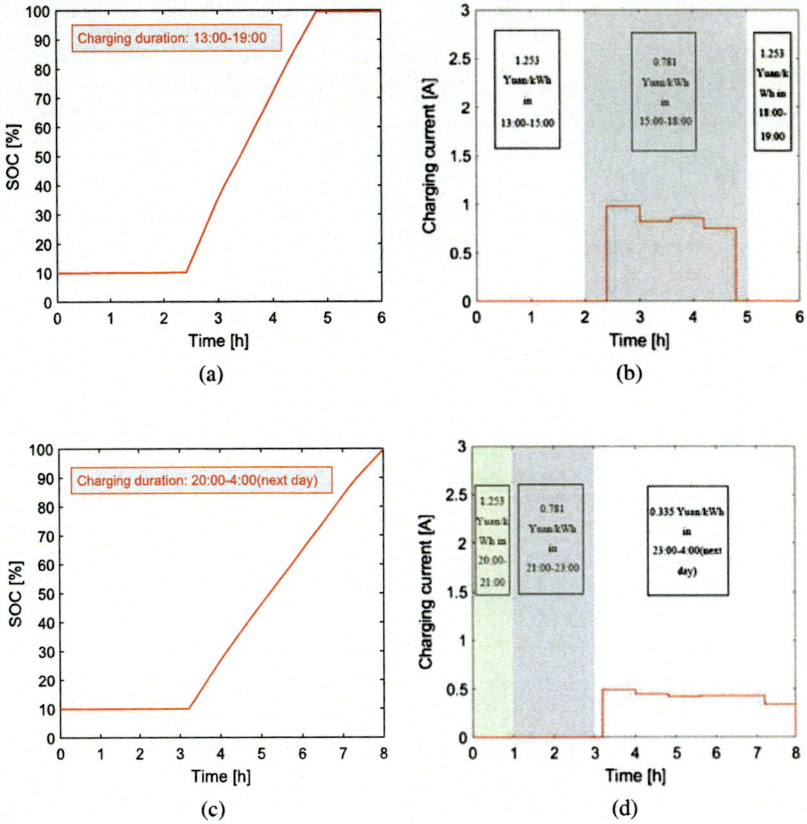

Fig. 6.9 **a** SOC response and **b** charging current for Case 4; **c** SOC response and **d** charging current for Case 5

Table 6.3 Simulation results for different user demands

	SOC gap (%)	Electricity fee (Yuan)	Energy loss (J)
Case 1	0.689	0.0055	972.93
Case 2	0.667	0.0087	555.33
Case 3	0.419	0.0055	634.25
Case 4	0.355	0.0054	262.58
Case 5	0.159	0.0023	140.42

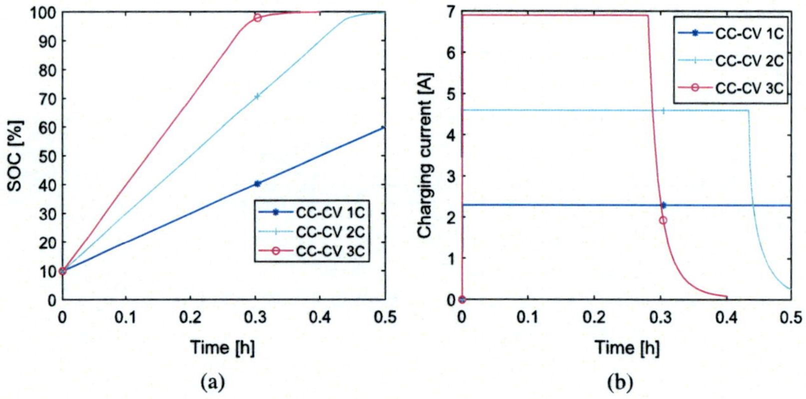

Fig. 6.10 **a** SOC responses, **b** charging currents under CC-CV control for the demand in Case 1

Table 6.4 Comparison results with CC-CV method

		SOC gap (%)	Electricity fee (Yuan)	Energy loss (J)
Case 1	CC-CV with 1 C-rate	40.03	0.003	287.53
	CC-CV with 2 C-rate	0.28	0.0056	1006.31
	CC-CV with 3 C-rate	0	0.0057	1339.39
	Proposed method	0.689	0.0055	972.93
Case 5	CC-CV with 1 C-rate	0	0.0088	605.12
	CC-CV with 2 C-rate	0	0.0089	1007.85
	CC-CV with 3 C-rate	0	0.0091	1339.39
	Proposed method	0.159	0.0023	140.42

6.5 Experimental Results

To further verify the performance of the designed user-involved charging control strategy with economic cost optimization, extensive experiments are conducted here on a Panasonic NCR18650B battery with a capacity of 2.85 Ah. Preliminary tests are first conducted to verify the model parameters of the battery. The mapping from the SOC to the open-circuit voltage of the battery is illustrated in Fig. 6.11a. The model parameters R_0, R_1, R_2, C_1, and C_2 are shown in Fig. 6.11b–f, respectively. The upper bounds of battery charging current, SOC, and terminal voltage are chosen as 2.85 A (1-C rate), 100%, and 4.2 V, respectively. The TOU electricity price is

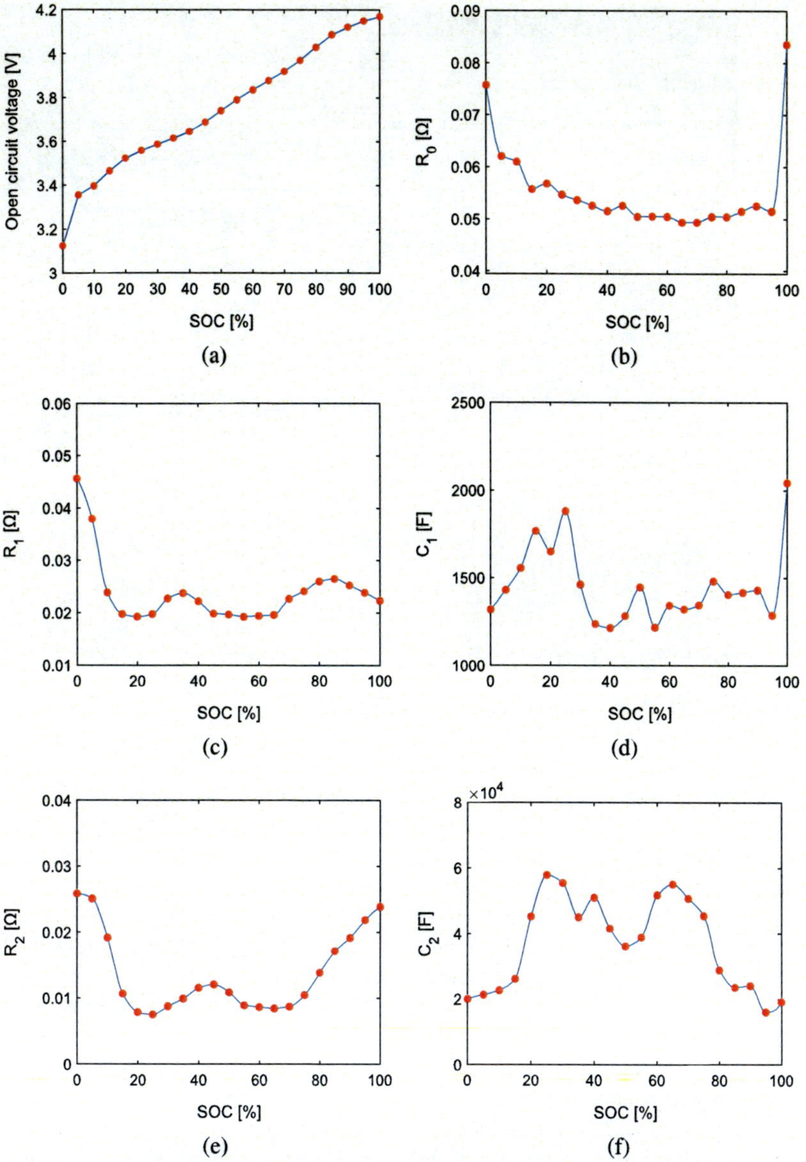

Fig. 6.11 **a** Open circuit voltage, **b** R_0, **c** R_1, **d** C_1, **e** R_2, and **f** C_2 in the battery model

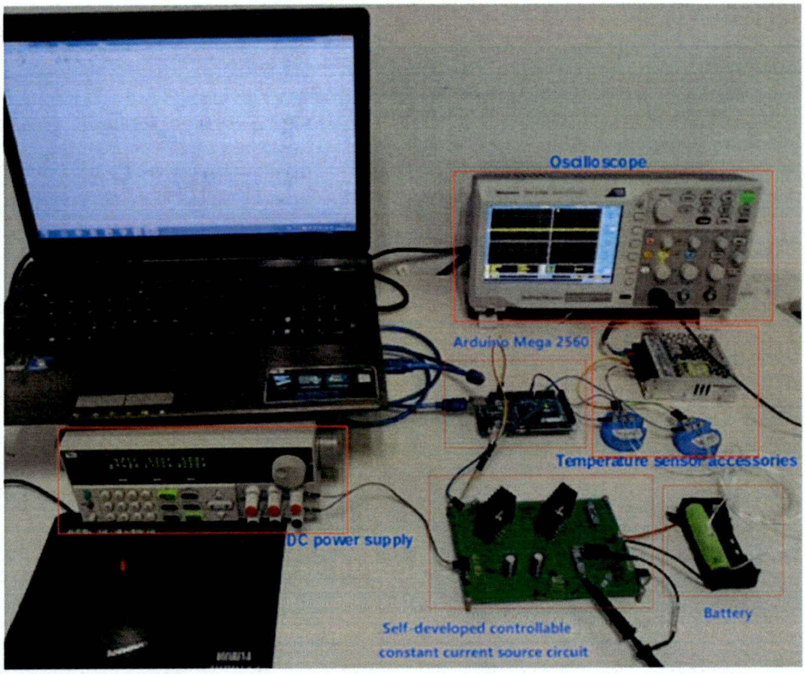

Fig. 6.12 Battery charging experimental platform

chosen as shown in Fig. 6.3. The experimental platform is shown in Fig. 6.12, where an Arduino Mega 2560 board controls a self-developed controllable constant current source circuit for battery charging. The experimental ambient temperature is about 19.5° C.

The target SOC is set to 100% and the desired charging duration is selected as 14:00–16:00. The experimental results in terms of the SOC, charging current, terminal voltage, and temperature of the battery are shown in Fig. 6.13a–d, respectively. From the above figure, it can be seen that under the proposed charging control strategy, the SOC of the battery reaches 99.53% after 2 h of charging, where the electricity cost and energy loss are 0.0088 Yuan and 1463.96 J, respectively. Specifically, it can be seen from Fig. 6.5 that a small charging current of about 0.6 A is used during 14:00–15:00 with an electricity price of 1.253 Yuan/kWh, while the maximum charging current during 15:00–16:00 is as high as 2.85 A with an electricity price of 0.781 Yuan/kWh, which is consistent with the previous analysis in this chapter.

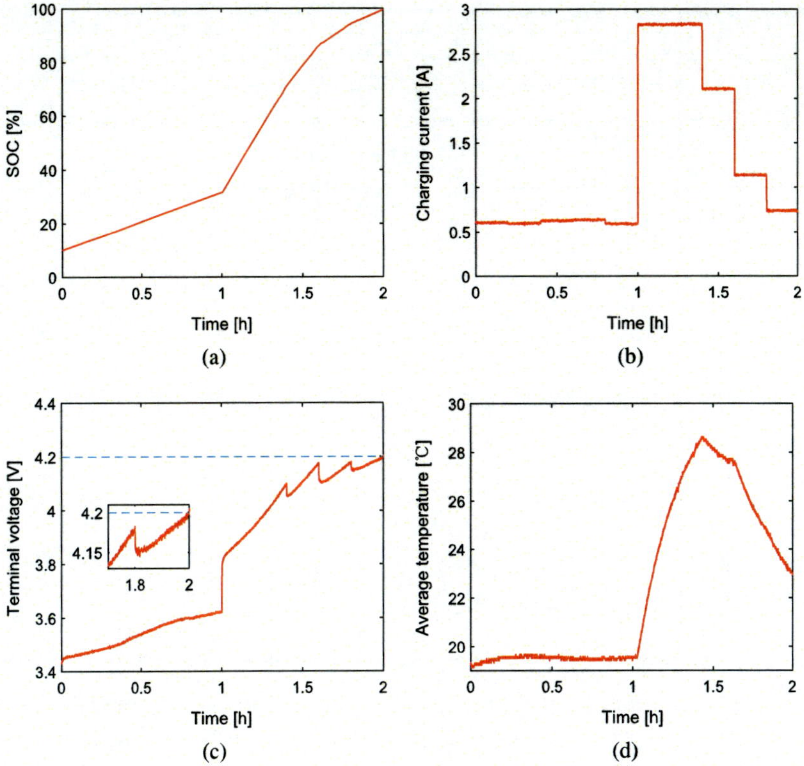

Fig. 6.13 Experimental results of the battery's **a** SOC, **b** current, **c** terminal voltage, and **d** temperature for the pre-set charging duration of 14:00–16:00

Finally, to further verify the effectiveness of the designed battery charging control strategy, experimental results for another user demand is provided in this chapter, where the charging duration is set to 21:00–1:00 (next day). As illustrated in Fig. 6.14, the results show that the proposed charging control method can indeed enable the battery to meet the charging target as well as the safety-related constraints.

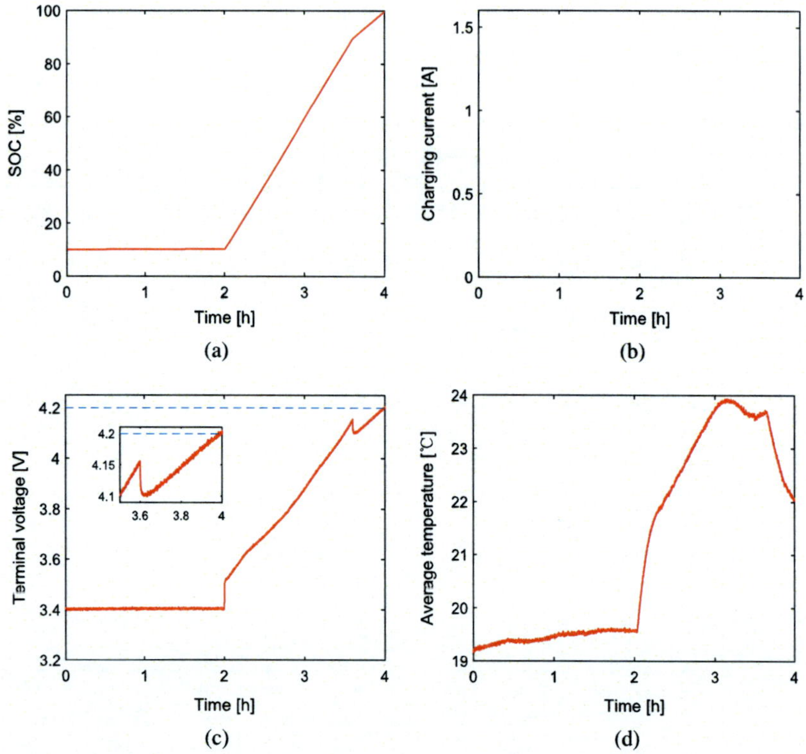

Fig. 6.14 Experimental results of the battery's **a** SOC, **b** current, **c** terminal voltage, and **d** temperature for the pre-set charging duration of 21:00–1:00 (next day)

References

1. Q. Ouyang, R. Fang, G. Xu, and Y. Liu, "User-involved charging control for EV lithium-ion batteries with economic cost optimization," *Applied Energy*, vol. 314, pp. 118878, 2022.
2. C. Zou, X. Hu, Z. Wei, and X. Tang, "Electrothermal dynamics-conscious lithium-ion battery cell-level charging management via state-monitored predictive control," *Energy*, vol. 141, pp. 250–259, 2017.
3. C. Chung, S. Jangra, Q. Lai, and X. Lin, "Optimization of electric vehicle charging for battery maintenance and degradation management," *IEEE Transactions on Transportation Electrification*, vol. 6, no. 3, pp. 958–969, 2020.
4. S. Su, H. Li, and D. W. Gao, "Optimal planning of charging for plug-in electric vehicles focusing on users' benefits," *Energies*, vol. 10, no. 7, pp. 952, 2017.
5. H. Min, W. Sun, X. Li, D. Guo, Y. Yu, T. Zhu, and Z. Zhao, "Research on the optimal charging strategy for Li-Ion batteries based on multi-objective optimization," *Energies*, vol. 10, no. 5, pp. 1–15, 2017.
6. S. Boyd, L. Vandenberghe, *Convex Optimization*, Cambridge University Press, New York, NY, USA, 2004.

7. X. Lin, H. E. Perez, S. Mohan, J. B. Siegel, A. G. Stefanopoulou, Y. Ding, and M. P. Castanier, "A lumped-parameter electro-thermal model for cylindrical batteries," *Journal of Power Sources*, vol. 257, pp. 1–11, 2014.
8. Y. Cao, S. Tang, C. Li, P. Zhang, Y. Tan, Z. Zhang, and J. Li, "An optimized EV charging model considering TOU price and SOC curve," *IEEE Transactions on Smart Grid*, vol. 3, no. 1, pp. 388–393, 2012.

Chapter 7
Charging Analysis for Lithium-Ion Battery Packs

Most existing charging efforts have focused on individual cells, and research on charge control of battery packs, which are more common in real-world applications, is not yet widespread. For example, in practical engineering, hundreds of batteries are usually connected in series as battery packs to provide the necessary high operating voltage of EVs. The battery pack in the Tesla vehicle is as illustrated in Fig. 7.1, which consists of 7000 battery cells. A major challenge of the current research related to battery pack charging control is the SOC imbalance among the cells. This challenge leads to the fact that once any cell in the pack reaches the upper threshold of SOC, the charging process of the whole pack must be terminated to avoid overcharging, even if the other cells have not been fully charged. In other words, the SOC imbalance phenomenon in the battery pack will limit its available capacity [1]. In summary, the cell SOC imbalance phenomenon makes charging control of a battery pack more complex than controlling a single cell because it is necessary to ensure that all cells are in balance during the charging process to maximize the effective capacity of the pack. In this Chapter, the charging problem is analysed for the battery pack and two battery pack chargers that can achieve cell equalization of the battery pack are introduced.

7.1 Cell Equalization Analysis

Due to its potential limitation of cathode and anode electrodes, a single lithium-ion battery cell's voltage is limited within the range of 2.5–4.2 V, which is obviously not sufficient to meet the EVs' high voltage requirement. Therefore, plenty of battery cells are usually connected in series as a high-voltage battery pack in practical EV applications. However, due to the inefficient manufacturing technology of batteries and the spatially uneven distribution of temperature within a battery pack, such a battery pack usually experiences cell imbalance due to slight inconsistency in each cell's characteristics.

Q. Ouyang and J. Chen, *Advanced Model-Based Charging Control for Lithium-Ion Batteries*, https://doi.org/10.1007/978-981-19-7059-7_7

Fig. 7.1 Battery pack in Tesla

In the battery pack, all cells share the same external charging/discharging current due to their serial connection. Since overcharging/overdischarging can shorten the battery cells' performance and lifetime, the charging/discharging process must be terminated when one of the cells is fully charged/discharged. It causes that the weakest cell restricts the performance of the battery pack, resulting in insufficient energy utilization of the entire battery pack [2]. Moreover, the cell inconsistency can accelerate battery degradation and even cause thermal runaway accidents. This illustrates the importance and necessity of cell equalization in the battery pack.

An extreme example is shown in Fig. 7.2, where the SOCs of the first and n-th cells in the serially connected battery pack are 0 and 100%, respectively. In this case, charging or discharging the battery pack can both cause damage to cell n or 1. Although the other cells still have a lot of remaining energy available, the entire battery pack cannot be used for safety protection. After the battery pack is balanced, the SOC of each cell is adjusted to the same level, and the cells in the battery pack can be fully charged or discharged at the same time, thus improving the effective capacity of the entire battery pack.

Conventional chargers typically charge all cells in a battery pack with the same current, which lacks the ability of mitigating the cell imbalance. If they are applied on the battery pack, the chargers would force termination of the entire pack charging process when any cell in the pack reaches the upper SOC threshold to avoid its overcharging, even though other cells are not fully charged. This will result in a significantly conservative use of the battery pack. Hence, we should construct a unique charging system for battery packs to achieve cell equalization. Here, two battery pack chargers are introduced: multi-module charger and traditional charger combined with equalizers.

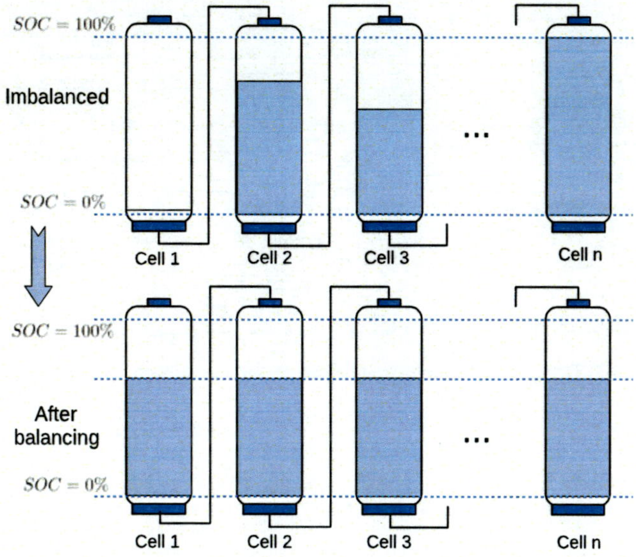

Fig. 7.2 Cell's SOCs before and after balancing

7.2 Multi-module Battery Pack Charger

In this section we consider a general charging setup that uses various power sources, e.g., alternating current (AC) grid, photovoltaic arrays, and local energy storage, to charge the battery pack. As shown in Fig. 7.3, a multi-module charger is designed for a battery pack consisting of n serially connected cells, where n modified isolated buck converter modules are used to charge each cell in the battery pack. This battery pack charger based on modified isolated buck converters has the advantage of being easy to implement, size-appropriate and cost-limited, as well as allowing for integrated infrastructure and modular design for easy retrofitting. In addition, the rapid development of integrated circuit technology can significantly reduce the production cost of battery chargers and facilitate the widespread production and use of the designed multi-module charger.

7.2.1 Model and Control of Battery Pack Charger

Figure 7.3 shows the modified isolated buck converter, which consists of a transformer T_i, an inductor L_i, a MOSFET Q_i, and two diodes d_{i1} and d_{i2}. Compared to the conventional buck converter, this converter has the advantage that it does not require the capacitor at the output port. The model equation for the i-th modified isolated buck converter with continuous conduction mode is

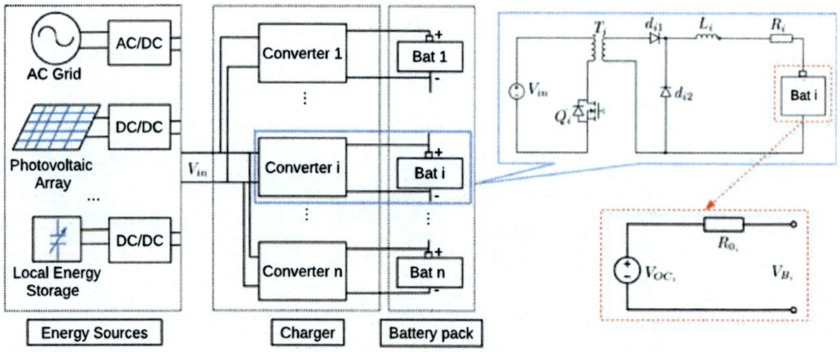

Fig. 7.3 Diagram of the multi-module charger for a serially connected lithium-ion battery pack

$$\frac{dI_{B_i}}{dt} = \frac{D_i}{L_i}V_{in} - \frac{V_{B_i}}{L_i} - \frac{I_{B_i}R_i}{L_i} \tag{7.1}$$

where V_{in} is the rectified DC input voltage, R_i represents the resistor for current detection, V_{B_i} and I_{B_i} denote the terminal voltage and charging current of the ith cell, and D_i $(0 \leq D_i \leq 1)$ is the duty cycle of the pulse width modulation (PWM) signal applied to the MOSFET Q_i. According to (7.1), the model of the multi-module charger can be expressed as follows:

$$\dot{v} = -Av + C_1u - C_2y \tag{7.2}$$

with

$$\begin{aligned} A &= diag\{\tfrac{R_1}{L_1}, \cdots, \tfrac{R_n}{L_n}\} \in \mathbb{R}^{n\times n} \\ C_1 &= diag\{\tfrac{V_{in}}{L_1}, \cdots, \tfrac{V_{in}}{L_n}\} \in \mathbb{R}^{n\times n} \\ C_2 &= diag\{\tfrac{1}{L_1}, \cdots, \tfrac{1}{L_n}\} \in \mathbb{R}^{n\times n} \end{aligned}$$

where $u = [u_1, \cdots, u_n]^T \triangleq [D_1, \cdots, D_n]^T \in \mathbb{R}^n$ is the control variable. In the designed multi-module charger, each cell in the battery pack is equipped with a modified isolated buck converter for independent charging, which can bring the following two benefits:

- Firstly, the charging control can be performed to the individual cell level, giving us more control authority over the charging procedure.
- Secondly, with the ability to control individual cells, we can finely tune the charging pattern for each cell to balance their SOCs to the same level.

The goal of the charger controller is to make the actual charging current through the multi-module charger track its desired value $I_{B_i}^d$ $(1 \leq i \leq n)$. From the energy-based perspective [3], we can define the Hamiltonian function of the charging process, which represents the total energy stored in the system as:

$$H(v) = \tfrac{1}{2} v^T C v \tag{7.3}$$

with $C = diag\{L_1, \cdots, L_n\} \in \mathbb{R}^{n \times n}$. Based on (7.2) and (7.3), the model of the charger can be represented as:

$$\dot{v} = -A_1 \tfrac{\partial H(v)}{\partial v} + C_1 u - C_2 y \tag{7.4}$$

with $A_1 = diag\{\frac{R_1}{L_1^2}, \cdots, \frac{R_n}{L_n^2}\} \in \mathbb{R}^{n \times n}$, where y is considered as a measurable disturbance vector. To ensure the charging currents v_i $(1 \leq i \leq n)$ tracking their desired values $\bar{I}^d_{B_i}$, an interconnection and damping assignment-passivity-based controller (IDA-PBC) is designed here for the system (7.4). It assigns a desired energy function with a minimum value at the desired equilibrium point by modifying the control input. The desired total energy function are defined as follows:

$$H_d(v) = \tfrac{1}{2}(v - v^d)^T C (v - v^d) \tag{7.5}$$

where $v^d = [bar I^d_{B_1}, \cdots, bar I^d_{B_n}]^T \in \mathbb{R}^n$. Since the required charging current is a slowly varying signal compared to the fast dynamics of the converter, the derivative of the required charging current can be assumed to be zero, i.e., $\dot{v}^d = 0_n$. It can be seen from (7.5) that $H_d(v) \geq 0$ and $H_d(v) = 0$ when and only when $v = v^d$, as C is positive definite. Suppose that a control input $u = \beta(v, v^d)$ can be found such that the closed-loop system of (7.4) satisfies

$$\dot{v} = -R_d \tfrac{\partial H_d(v)}{\partial v} \tag{7.6}$$

where $R_d = diag\{R_{d_1}, \cdots, R_{d_n}\} \in \mathbb{R}^{n \times n}$ is a designed positive definite matrix. According to (7.4) and (7.6), it generates

$$-R_d \tfrac{\partial H_d(v)}{\partial v} = -A_1 \tfrac{\partial H(v)}{\partial v} + C_1 \beta(v, v^d) - C_2 y. \tag{7.7}$$

The control input can be inferred from (7.7) that

$$\begin{aligned} u &= \beta(v, v^d) \\ &= C_1^{-1}((A_1 - R_d)Cv + R_d C v^d + C_2 y) \end{aligned} \tag{7.8}$$

where C_1^{-1} is the inverse matrix of C_1.

Convergence proof: We select a Lyapunov function is selected as follows:

$$V = H_d(v) \tag{7.9}$$

where $H_d(v)$ can be seen in (7.5). According to (7.3)–(7.8), the derivative of (7.9) satisfies

$$\dot{V} = \frac{\partial H_d(v)}{\partial v}\dot{v} \leq -P\tilde{v}^2 \tag{7.10}$$

where $P = CR_dC$ and $\tilde{v} = (v - v^d)$. Since P is positive definite, $\dot{V} \leq 0$ and $\dot{V} = 0$ when and only when $v = v^d$. Based on the LaSalle's invariance principle [4], it yields that $v \to v^d$. Thus, the convergence of the charging current tracking system is proven.

As the modified buck converters are actually controlled independently, the controller (7.8) can be rewritten as n distributed IDA-PBCs as follows:

$$u_i = \frac{(R_i - R_{d_i}L_i^2)v_i + R_{d_i}L_i^2 v_i^d + V_{B_i}}{V_{in}} \tag{7.11}$$

for $1 \leq i \leq n$, where R_{d_i} is chosen to maintain the duty cycle u_i in the set [0, 1].

7.2.2 Performance Validation

As shown in Fig. 7.4a, a modified multi-module charger based on the isolated buck converter is modeled in dSPACE by real-time simulation with inductor $L_i = 0.01$ H ($1 \leq i \leq 3$), non-ideal transformer T_i, MOSFET Q_i with internal diode resistors 0.1 Ω and FET resistors 0.01 Ω, and snubber resistors 500 Ω for diodes d_{i1} and d_{i2}. The DC input voltage is $V_{in} = 24$ V and the frequency of the PWM signal applied to the MOSFET is set to 5 KHz. The current sense resistor is selected as $R_i = 1$ Ω. The performance of the modified isolated buck converter is verified by experiments. The expected and actual charging currents of the modified buck converter with IDA-PBC algorithm are shown in Fig. 7.4, where the actual current is obtained by measuring the voltage of the current sense resistor through the Keysight oscilloscope. The

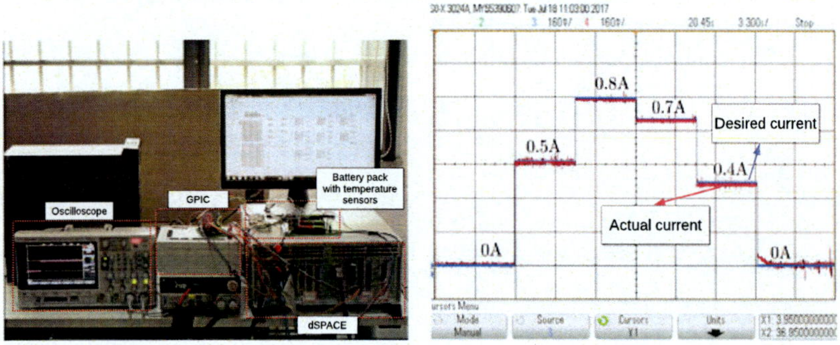

(a) (b)

Fig. 7.4 **a** Experimental test bench, **b** expected and actual current responses of the modified buck converter with IDA-PBC algorithm experimental results show that the designed IDA-PBC algorithm can make the actual charging current through the converter well track the expected value.

7.3 Battery Pack Charging System Combining Traditional Charger and Equalizers

It is important to note that the multi-module charger mentioned above may increase costs compared to conventional battery chargers, especially for large battery packs, because the designed charger consists of many small charger modules. Another choice of the battery pack charging system is the integration of a conventional charger and cell equalizers. It can not only effectively solve the problem of cell imbalance, but also avoid the cost increase of the battery charger in practical applications, since the battery equalization system is usually equipped as an important part of the BMS [1]. Here, we will introduce the commonly utilized cell equalization system and then describe the hardware of two types of equalizers.

7.3.1 Classification of Equalization Systems

Current cell balance methods can be divided into two categories: passive and active, which are also known as dissipative and non-dissipative, respectively.

- Passive cell balancing methods utilize the shunt resistors to dissipate the excess charge of high-SOC cells, in an attempt to match those with lower SOC. Despite its advantages such as simplicity, low price, and reliability, the dissipation of energy in passive cell balancing methods leads to a reduction in the battery pack's capacity and a need for an additional thermal management system.
- During active cell balancing methods, the charge is spread from cells with higher SOC to those with lower SOC using active equalizing circuits. Unlike passive equalizing strategies that waste energy, active cell balancing enables energy to be transmitted between cells, making it more energy-efficient and requiring less equalization time.

In recent years, researchers have been deeply interested in active cell balancing strategies because of the above mentioned advantages. As displayed in Fig. 7.5, the active cell balancing systems, based on energy transfer, are mainly divided into adjacent-based, non-adjacent-based, direct cell-cell, and mixed topologies [5]. Following items are descriptions of these topologies that are further introduced.

- Adjacent-based topology utilizes active equalizers performed on the adjacent cells or modules of a battery pack, to measure the SOC or average SOC level and balance it among the whole battery pack, according to a definite energy transferring strategy from the high cell or module SOC level to the lower one. These kinds of balancing

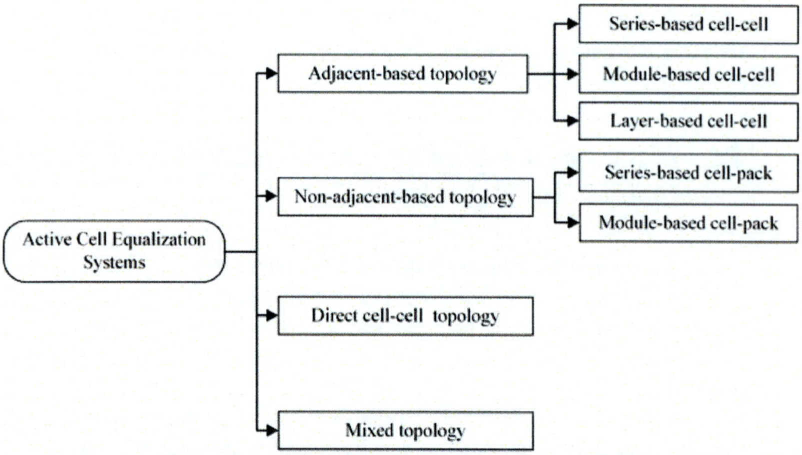

Fig. 7.5 Classification of active equalization systems [5]

topologies are widely used in the studying battery balancing applications, and regarding the topology-level structure could be further categorized as series-based cell-cell (as shown in Fig. 7.6), module-based cell-cell (as shown in Fig. 7.7), and layer-based cell-cell topologies (as shown in Fig. 7.8).

- Non-adjacent-based topology employs active equalizers in connection with each cell or module and the whole battery pack, to measure and balance the SOC levels of the cells or modules based on a defined energy flow strategy that transfers the charge between cells or modules and pack. Subsequently, the non-adjacent-based topology is divided into series-based cell-pack (as shown in Fig. 7.9) and module-based cell-pack (as shown in Fig. 7.10).
- Direct cell-cell topology (as depicted in Fig. 7.11) is applied by utilizing a typical equalizer by means of a couple of switches for each cell and other electrical components such as a capacitor, to actively transfer the energy from any cell to any cell by using a proper control scheme, regardless of whether the cells are adjacent or non-adjacent. For each typical working time, only the switches connected with the cell whose SOC has the largest bias with the average SOC of the pack can be closed to avoid the short circuit. Consequently, the cells' SOC level of the whole pack reaches the same level after continuously connecting the switches associated with those individual cells.
- Mixed topology is equipped with two complementary balancing layers in which the adjacent-based or non-adjacent-based topologies are utilized in each layer, as illustrated in Fig. 7.12. Subsequently, the battery pack is divided into several modules that each module involves the same number of cells. The bottom layer takes the balancing of the cells within each module through the applied topology under the control of a slave BMS. Subsequently, the top layer is responsible for balancing the charge between the modules through the associated topology under

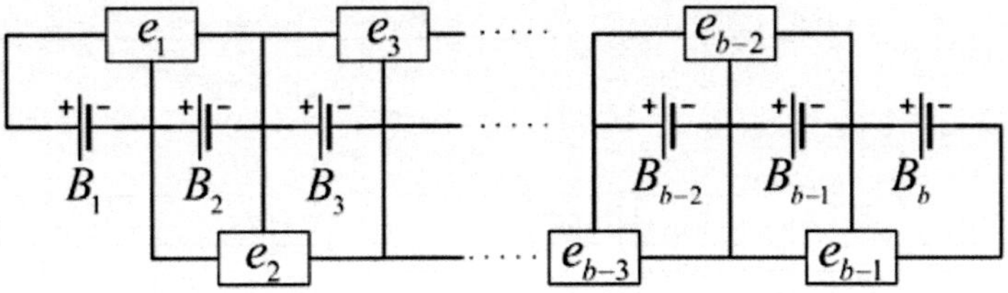

Fig. 7.6 Series-based cell-cell equalization

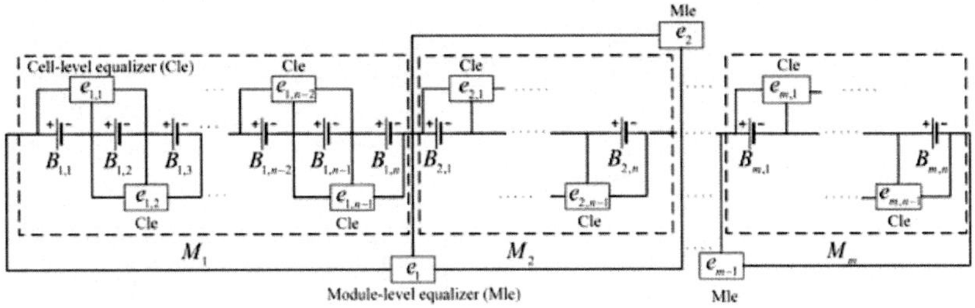

Fig. 7.7 Module-based cell-cell equalization

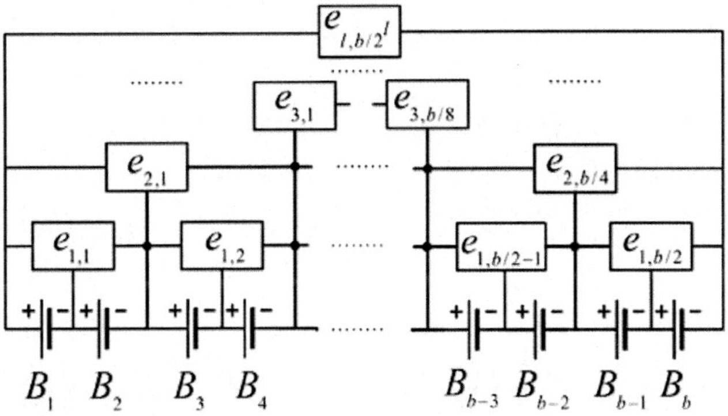

Fig. 7.8 Layer-based cell-cell equalization

control of a master BMS. The slave and master BMS are in connection with each other, and apparently, this topology can provide integration of both top and bottom layer topology merits.

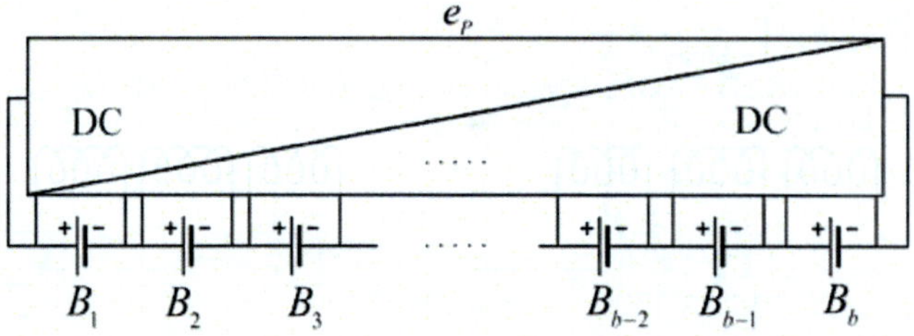

Fig. 7.9 Series-based cell-pack equalization

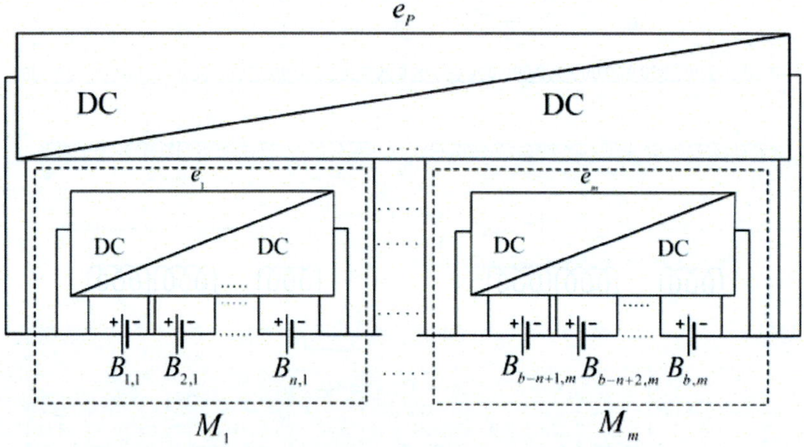

Fig. 7.10 Module-based cell-pack equalization

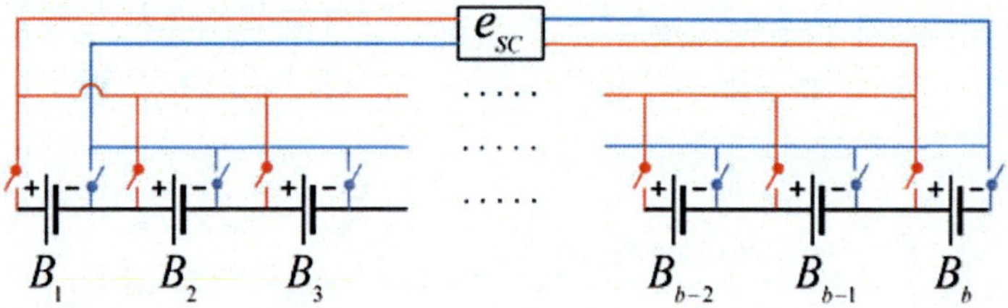

Fig. 7.11 Direct cell-cell equalization

7.3.2 Bidirectional Modified Cûk Converter-Based Equalizer

The bidirectional modified Cûk converters [6] are commonly utilized as cell balancing circuits. Battery cell i and cell $i + 1$ $(1 \leq i \leq n - 1)$ are connected through the i-th converter, as illustrated in Fig. 7.13. The converter consists of two uncoupled inductors L_{i1} and L_{i2}, an energy transferring capacitor C_i, and two MOSFETs Q_{i1}

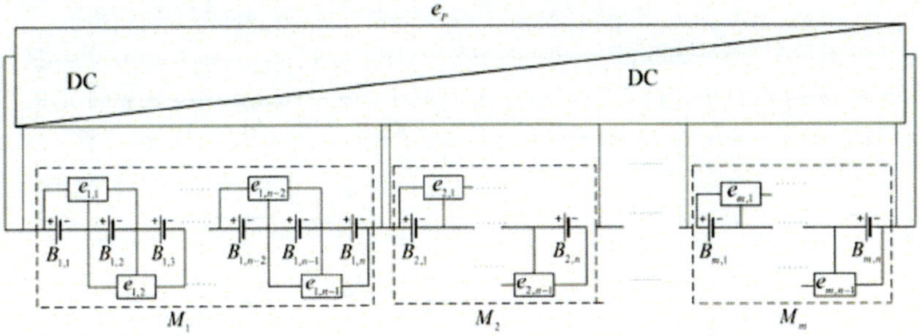

Fig. 7.12 Mixed equalization topology with series-based cell-cell in bottom layer and module-based cell-pack in the top layer

and Q_{i2} with body diodes d_{i1} and d_{i2}. The circuit is driven by PWM signals that control MOSFETs to be on or off. The duty cycles of the PWM signals are chosen as the control variables for cell equalizing. For the discontinuous inductor current mode (DICM) operation, the MOSFETs in the converters are turned on and the body diodes are turned off at the zero current state, which will significantly reduce the power loss of the MOSFET switch and improve the average balancing efficiency compared with those designed in continuous inductor current mode [7]. Hence, the converters are designed to operate in DICM to reduce the energy losses during the switching of MOSFETs by properly selecting the required duty cycles.

The i-th converter can transfer energy between its connected cell i and cell $i+1$ in bidirectional directions, which has a symmetrical structure. For the DICM operation, as illustrated in Fig. 7.14, the dynamic response of the i-th converter when the energy is transferred from the i-th/$(i+1)$-th cell to the $(i+1)$-th/i-th cell can be divided into three states in one switching period T_s. Without loss of generality, we take the situation that energy is transferred from the i-th cell to the $(i+1)$-th cell as an example to analyse the converter model as follows:

State 1: The initial ith capacitor voltage $v_{C_i}(t)$ equals $V_{B_i} + V_{B_{i+1}}$. The initial inductor currents $i_{L_{i1}}(t)$ and $i_{L_{i2}}(t)$ are both zero. During the time period $[0, D_{i1}T_s)$, MOSFET Q_{i1} is turned on. The energy stored in the capacitor C_i is transferred to the $(i+1)$-th cell and the inductor L_{i1} stores energy from the ith cell. The dynamics can be expressed as

$$\begin{bmatrix} \frac{di_{L_{i1}}}{dt} \\ \frac{di_{L_{i2}}}{dt} \\ \frac{dv_{C_i}}{dt} \end{bmatrix} = \begin{bmatrix} 0 & 0 & 0 \\ 0 & 0 & \frac{1}{L_{i2}} \\ 0 & \frac{-1}{C_i} & 0 \end{bmatrix} \begin{bmatrix} i_{L_{i1}} \\ i_{L_{i2}} \\ v_{C_i} \end{bmatrix} + \begin{bmatrix} \frac{1}{L_{i1}} & 0 \\ 0 & \frac{-1}{L_{i2}} \\ 0 & 0 \end{bmatrix} \begin{bmatrix} V_{B_i} \\ V_{B_{i+1}} \end{bmatrix} \tag{7.12}$$

where $i_{L_{i1}}(t)$ and $i_{L_{i2}}(t)$ denote the current of L_{i1} and L_{i2}, respectively; V_{B_i} and $V_{B_{i+1}}$ are the terminal voltages of the i-th and $(i+1)$-th cell; D_{i1} denotes the duty cycle of the PWM signal applied on MOSFET Q_{i1}.

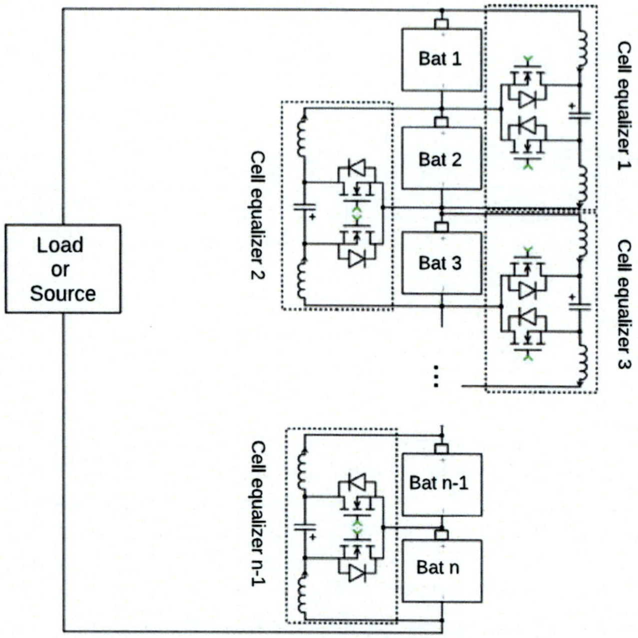

Fig. 7.13 Schematic of the bidirectional modified Cûk converter

State 2: During the time period $[D_{i1}T_s, (D_{i1} + D'_{i1})T_s)$, MOSFET Q_{i1} is turned off and diode d_{i2} is forced to turn on. The capacitor C_i is charged by the i-th cell and the stored energy in L_{i2} is transfered to the $(i + 1)$-th cell. The dynamic equations is obtained as follows:

$$\begin{bmatrix} \frac{di_{L_{i1}}}{dt} \\ \frac{di_{L_{i2}}}{dt} \\ \frac{dv_{C_i}}{dt} \end{bmatrix} = \begin{bmatrix} 0 & 0 & \frac{-1}{L_{i1}} \\ 0 & 0 & 0 \\ \frac{1}{C_i} & 0 & 0 \end{bmatrix} \begin{bmatrix} i_{L_{i1}} \\ i_{L_{i2}} \\ v_{C_i} \end{bmatrix} + \begin{bmatrix} \frac{1}{L_{i1}} & 0 \\ 0 & \frac{-1}{L_{i2}} \\ 0 & 0 \end{bmatrix} \begin{bmatrix} V_{B_i} \\ V_{B_{i+1}} \end{bmatrix} \tag{7.13}$$

where $(D_{i1} + D'_{i1})T_s$ is the time that the inductor currents decrease to zero.

State 3: During the time period $[(D_{i1} + D'_{i1})T_s, T_s)$, the MOSFET Q_{i1} and diode d_{i2} are both turned off. There is no charge transferred between the i-th and $(i + 1)$-th cells. It can be obtained that

$$\begin{aligned} i_{L_{i1}} &= 0 \\ i_{L_{i2}} &= 0 \\ v_{C_i} &= V_{B_i} + V_{B_{i+1}}. \end{aligned} \tag{7.14}$$

According to the principle of inductor volt-second balance and the principle of capacitor charge balance, for a time period T_s, the average inductor currents $I_{L_{i1}}$, $I_{L_{i2}}$

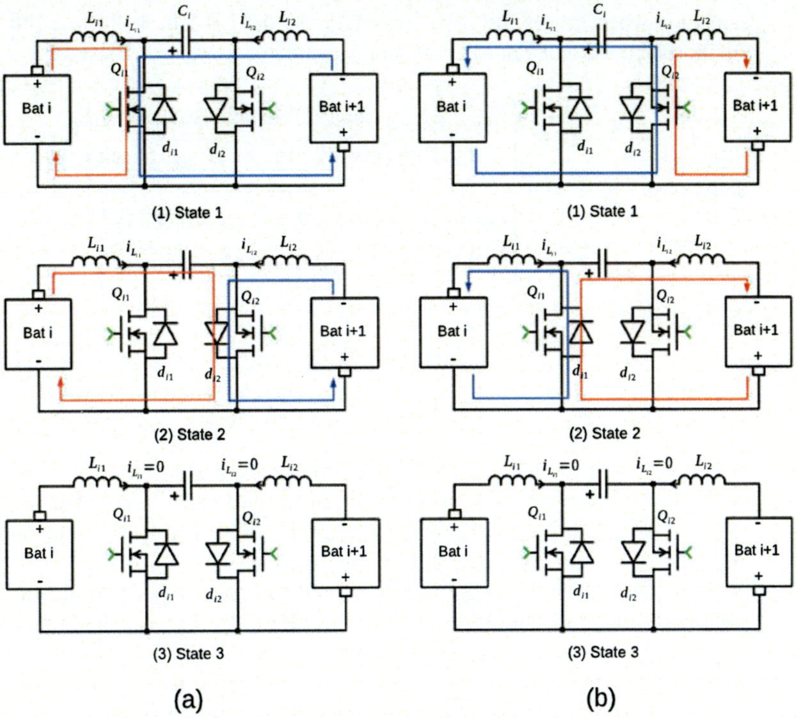

Fig. 7.14 Dynamics of the modified Cûk converter when energy is transferred **a** from cells i to $i+1$, **b** from cells $i+1$ to i

and the average capacitor voltage V_{C_i} keep as constants when the converter operates in a steady state. Based on (7.12) and (7.13), it can be obtained that

$$\frac{D'_{i1}T_s(V_{C_i}-V_{B_i})}{L_{i1}} = \frac{D_{i1}T_sV_{B_i}}{L_{i1}}. \tag{7.15}$$

The average currents of L_{i1} and L_{i2} are deduced as follows:

$$\begin{aligned} I_{L_{i1}} &= f_{i1}(D_{i1}) = \frac{1}{2}\frac{D_{i1}T_sV_{B_i}(D_{i1}+D'_{i1})}{L_{i1}} \\ I_{L_{i2}} &= f_{i2}(D_{i1}) = \frac{1}{2}\frac{D_{i1}T_s(V_{C_i}-V_{B_{i+1}})(D_{i1}+D'_{i1})}{L_{i2}} \end{aligned} \tag{7.16}$$

where the average inductor currents $I_{L_{i1}}$, $I_{L_{i2}}$ and average capacitor voltage V_{C_i} are constants. Similarly, the average currents of L_{i2} and L_{i1} can also be deduced for the situation where the energy is transferred from the $i+1$-th cell to the i-th cell. To unify the notation, it can be expressed as

$$
\begin{aligned}
I_{L_{i1}} &= \begin{cases} f_{i1}(D_{i1}), & \text{energy from cell } i \text{ to cell } i+1 \\ p_i' f_{i2}(D_{i2}), & \text{energy from cell } i+1 \text{ to cell } i \end{cases} \\
I_{L_{i2}} &= \begin{cases} p_i f_{i1}(D_{i1}), & \text{energy from cell } i \text{ to cell } i+1 \\ f_{i2}(D_{i2}), & \text{energy from cell } i+1 \text{ to cell } i \end{cases}
\end{aligned} \tag{7.17}
$$

with

$$
\begin{aligned}
f_{i1}(D_{i1}) &= \frac{1}{2}\frac{T_s V_{B_i} V_{C_i} D_{i1}^2}{L_{i1}(V_{C_i}-V_{B_i})} \\
f_{i2}(D_{i2}) &= -\frac{1}{2}\frac{T_s V_{B_{i+1}} V_{C_i} D_{i2}^2}{L_{i2}(V_{C_i}-V_{B_{i+1}})}
\end{aligned} \tag{7.18}
$$

where p_i and p_i' are the energy transfer efficiencies, which satisfies $0 \le p_i \le 1$ and $0 \le p_i' \le 1$. The current direction out of the i-th cell and the current direction into the $i+1$-th cell are defined as the reference direction of $I_{L_{i1}}$ and $I_{L_{i2}}$, respectively. Referring to [7], V_{C_i} can be obtained by approximating $V_{C_i} \approx V_{B_i} + V_{B_{i+1}}$. The negative sign of $f_{i2}(D_{i2})$ in (7.18) indicates that its direction is opposite to the pre-defined reference direction. The average inductor current $I_{L_{i1}}/I_{L_{i2}}$ is called the controlled equalizing current here, when the energy is transferred from the i-th/$i+1$-th cell to the $i+1$-th/i-th cell. Based on (7.17) and (7.18), for a typical controlled equalizing current $I_{L_{i1}}/I_{L_{i2}}$, the corresponding duty cycles are determined as:

$$
\begin{aligned}
D_{i1} &= \begin{cases} \sqrt{\frac{2L_{i1}(V_{C_i}-V_{B_i})I_{L_{i1}}}{T_s V_{B_i} V_{C_i}}}, & \text{energy from cells } i \text{ to } i+1 \\ 0, & \text{energy from cells } i+1 \text{ to } i \end{cases} \\
D_{i2} &= \begin{cases} 0, & \text{energy from cells } i \text{ to } i+1 \\ \sqrt{-\frac{2L_{i2}(V_{C_i}-V_{B_{i+1}})I_{L_{i2}}}{T_s V_{B_{i+1}} V_{C_i}}}, & \text{energy from cells } i+1 \text{ to } i. \end{cases}
\end{aligned} \tag{7.19}
$$

It ensures that the modified Cûk converter transports energy from one cell to its adjacent cell only in a particular direction at the same time period.

Performance validation: The test bench including three self-developed bidirectional modified Ćuk converters is shown in Fig. 7.15. The i-th ($1 \le i \le 3$) converter's parameters are selected as $L_{i1} = L_{i2} = 100\ \mu H$ and $C_i = 470\ \mu F$. NTD6416AN-1G MOSFETs are driven by the PWM signals with a 7 kHz switching frequency. A preliminary experiment is conducted to validate the performance of one bidirectional modified Ćuk converter connecting two cells. Its connected cells' terminal voltages are $V_{B_1} = 3.93$ V and $V_{B_2} = 3.62$ V, respectively. The duty cycle of the PWM signal applied on MOSFET Q_1 is set to 0.3. The transient inductor current curves recorded by a Keysight oscilloscope are denoted in Fig. 7.16, which are obtained by measuring

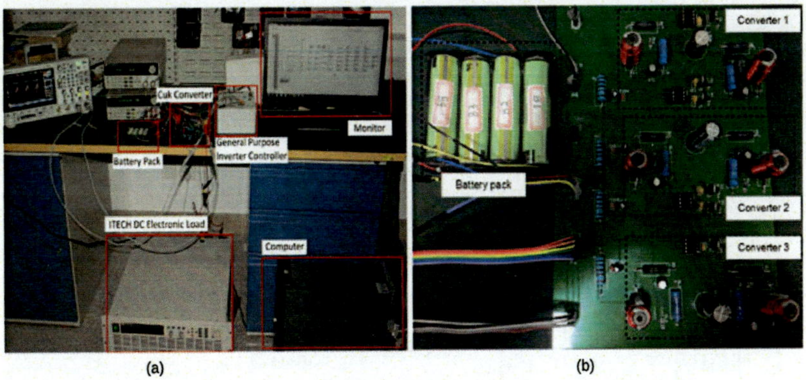

Fig. 7.15 **a** Experimental test bench, **b** bidirectional modified Ćuk converter board

Fig. 7.16 Inductor currents of the modified Ćuk converter

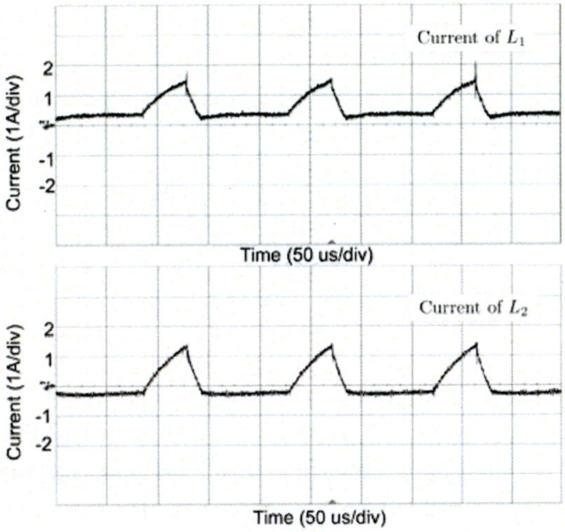

the resistors' voltages serially connected with the inductors. The average inductor currents are $I_{L_1} = 0.147$ A and $I_{L_2} = 0.129$ A, with the energy transfer efficiency of 0.878.

7.3.3 Modified Isolated Bidirectional Buck-Boost Converter-Based Equalizer

Another commonly used equalizer is the modified isolated bidirectional buck-boost converter [8], whose operational principle is shown as follows:

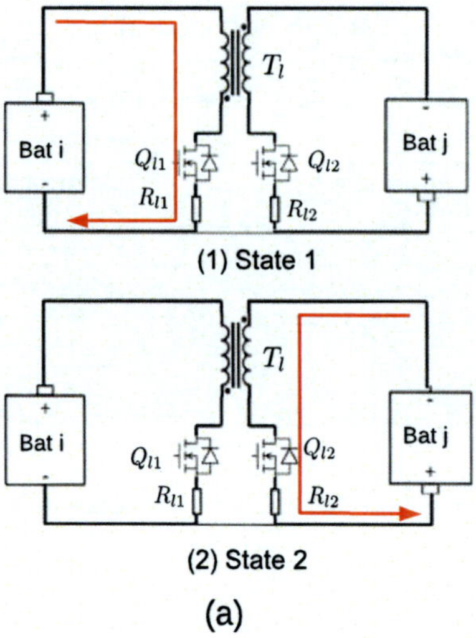

Fig. 7.17 Balancing of two cells by a modified isolated buck-boost converter

Operational principle: As shown in Fig. 7.17, the schematic diagram of two cells connected by a modified isolated bidirectional buck-boost equalizer, which is taken from an entire battery equalization system. The equalizer, numbered l, consists of two MOSFETs Q_{l1} and Q_{l2}, two resistors R_{l1} and R_{l2}, and a transformer T_l. Since it has a symmetric structure, it allows a bidirectional transfer of energy between the i-th and j-th cells it connects. When the energy is transferred from the i-th/j-th cell to the j-th/i-th cell, as illustrated in Fig. 7.17a, b, the balancing process can be divided into two alternating steps within one switching cycle:

- State 1: MOSFET Q_{l1}/Q_{l2} is turned on, and MOSFET Q_{l2}/Q_{l1} is turned off. Transformer T_l is charged by the i-th/j-th cell.
- State 2: MOSFET Q_{l1}/Q_{l2} is turned off, and MOSFET Q_{l2}/Q_{l1} is turned on. Transformer T_l charges the j-th/i-th cell.

By repeating the above two states, energy can be transferred from the i-th/j-th cell to the j-th/i-th cell. According to the inductive volt-second balance principle, the average inductance current at each step remains constant when the converter is operating in steady state. Therefore, the average inductance current can be used as the equalization current.

Performance validation: For the l-th modified isolated buck-boost converter connected to the i-th and j-th cells, the circuit parameters are chosen as $R_{l1} = R_{l2} = 25\ \Omega$ and $T_l = 10\ \mu$H, where the resistors are used to measure the equalization current. The voltages of cell i and cell j are set to $V_{B_i} = 3.91$ V and $V_{B_j} = 3.72$ V, respectively. By simulating the cell equalization procedure using Cadence Dspice

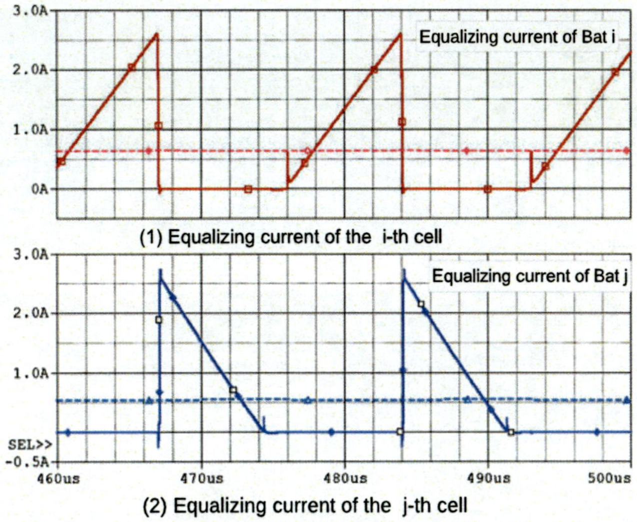

Fig. 7.18 Equalizing currents supplied by the i-th and to the j-th cells

tool, the transient equalization current curves through the isolated bidirectional buck-boost converter are shown in Fig. 7.18. Their average equalization currents are about 0.63 A and 0.54 A, respectively. The energy transfer efficiency through the l-th converter is about 85%. This agrees with the analysis of the modified isolated buck-boost converter.

In addition, the constructed experimental test bench is shown in Fig. 7.19, in which the self-developed modified isolated buck boost converter is illustrated in Fig. 7.19b. It consists of a WE-FB Flyback transformer of 3.46 μH, two NTD6416AN-1G MOSFETs, a MAX627CPA+ dual-power MOSFET driver, and two 0.5 Ω resistors.

An experiment is performed to validate the performance of an isolated bidirectional buck-boost converter connected to two cells. The PWM frequency is selected as 124 KHz, and the duty cycles applied on the MOSFETs Q_1 and Q_2 are set to 63% and 35%, respectively. The two connected cells' terminal voltages are $V_{B_1} = 3.82$ V and $V_{B_2} = 3.71$ V, respectively. The instantaneous equalization current curves of cells through the connected equalizer are shown in Fig. 7.20b, which are obtained by measuring the voltages across the resistors R_1 and R_2 in the converter. The average equalization currents are $I_1 = 0.1798$ A and $I_2 = 0.1084$ A, respectively.

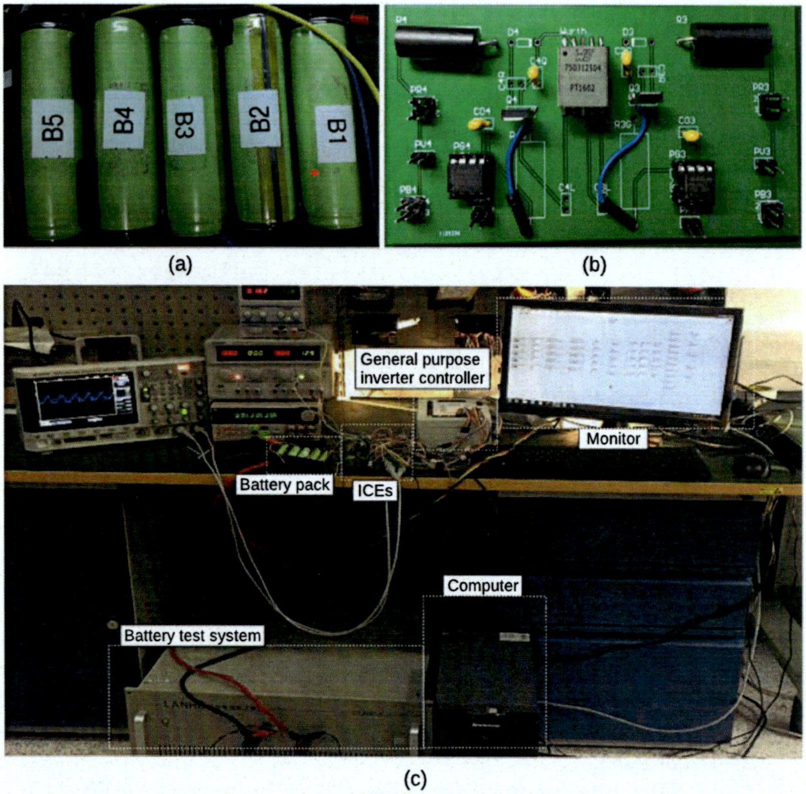

Fig. 7.19 **a** Battery pack, **b** isolated bidirectional buck-boost converter, **c** experimental test bench

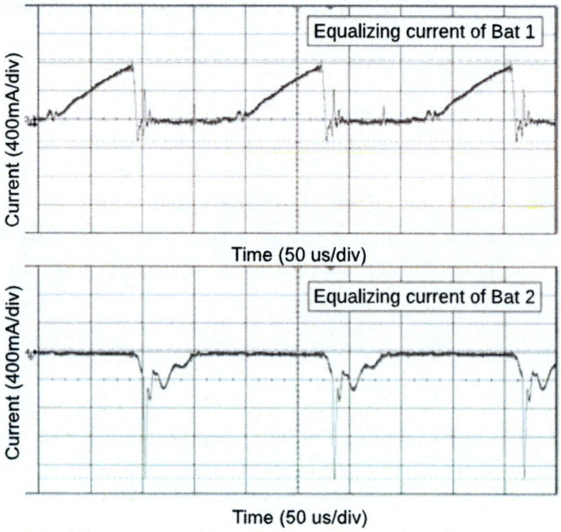

Fig. 7.20 Equalizing currents through the isolated bidirectional buck-boost converter

References

1. J. Gallardo-Lozano, E. Romero-Cadaval, M. I. Milanes-Montero, and M. A. Guerrero-Martinez, "Battery equalizing active methods," *Journal of Power Sources*, vol. 246, pp. 934–949, 2014.
2. Q. Ouyang, W. Han, C. Zou, G. Xu, and Z. Wang, "Cell balancing control for lithium-ion battery packs: A hierarchical optimal approach," *IEEE Transactions on Industrial Informatics*, vol. 16, no. 8, pp. 5065–5075, 2020.
3. J. Zeng, Z. Zhang, and W. Qiao, "An interconnection and damping assignment passivity-based controller for a DC-DC boost converter with a constant power load," *IEEE Transactions on Industrial Applications*, vol. 50, no. 4, pp. 2314–2322, 2014.
4. H. K. Khalil, *Nonlinear Systems*, 3rd ed. Englewood Cliffs, NJ, USA: Prentice-Hall, 2002.
5. N. Ghaeminezhad, Q. Ouyang, X. Hu, G. Xu, and Z. Wang, "Active cell equalization topologies analysis for battery packs: A systematic review," *IEEE Transactions on Power Electronics*, vol. 36, no. 8, pp. 9119–9135, 2021.
6. Q. Ouyang, J. Chen, J. Zheng, and Y. Hong, "SOC estimation-based quasi-sliding mode control for cell balancing in lithium-ion battery packs," *IEEE Transactions on Industrial Electronics*, vol. 65, no. 4, pp. 3427–3436, 2018.
7. Y. S. Lee, M. Chen, "Intelligent control battery equalization for series connected lithium-ion battery strings," *IEEE Transactions on Industrial Electronics*, vol. 52, no. 5, pp. 1297–1307, 2005.
8. Q. Ouyang, J. Chen, J. Zheng, and H. Fang, "Optimal cell-to-cell balancing topology design for serially connected lithium-ion battery packs," *IEEE Transactions on Sustainable Energy*, vol. 9, no. 1, pp. 350–360, 2018.

Chapter 8
User-Involved Charging Control for Battery Packs: Centralized Structure

To achieve cell equalization at the end of the charging procedure to enhance the effective usable capacity of the entire battery pack, a multi-module battery pack charger as shown in Fig. 7.3 is utilized. Based on this charger, an optimal multi-objective charging issue is formulated in this Chapter that takes account of the user demand, cell equalization, and temperature rise simultaneously, and then an optimal charging current scheduling strategy is derived, where a variable scheduling sampling period algorithm is utilized to improve the computational efficiency [1].

8.1 Battery Pack Model and Constraints

8.1.1 Battery Pack Model

This work takes into account the Rint model shown in Figs. 3.5 and 7.3 to describe the dynamics of each cell in a n-modular battery pack. This model is generally suitable for characterizing the battery operating within the normal temperature range and it can achieve a balance between model accuracy and computational complexity. As illustrated in Fig. 7.3, the battery model is represented by a voltage source and a resistor. The terminal voltage of the i-th cell can be calculated as follows:

$$V_{B_i} = V_{OC_i} + R_{0_i} I_{B_i} \tag{8.1}$$

where V_{B_i} and V_{OC_i} represent the terminal voltage and OCV of i-th cell in the battery pack, respectively; R_{0_i} and I_{B_i} denote the internal resistor and current of the i-th cell. The OCV and internal resistance of the i-th cell are nonlinear functions of its SOC, which can be expressed as:

Q. Ouyang and J. Chen, *Advanced Model-Based Charging Control for Lithium-Ion Batteries*, https://doi.org/10.1007/978-981-19-7059-7_8

$$V_{OC_i} = f_i(SOC_i), \quad R_{0_i} = h_i(SOC_i) \tag{8.2}$$

where SOC_i represents the i-th cell's SOC, whose dynamics can be obtained as:

$$\frac{\mathrm{d}SOC_i}{\mathrm{d}t} = \frac{\eta_i}{Q_i} I_{B_i} \tag{8.3}$$

where η_i is the Coulombic efficiency, and Q_i is the i-th cell's capacity in Ampere-hour. The i-th cell's SOC is assumed to be known in the charging control strategy design throughout this book since it can be estimated by the neural network-based nonlinear observer and robust observer as in Chaps. 4 and 5, respectively. Based on (8.1)–(8.3), the model of the n-modular battery pack can be rewritten as:

$$\begin{aligned} \dot{x} &= B_1 v \\ y &= f(x) + h(x)v \end{aligned} \tag{8.4}$$

with

$$\begin{aligned} B_1 &= diag\{\tfrac{\eta_1}{Q_1}, \ldots, \tfrac{\eta_n}{Q_n}\} \\ f(\cdot) &= [f_1(\cdot), \ldots, f_n(\cdot)]^T \\ h(\cdot) &= diag\{h_1(\cdot), \ldots, h_n(\cdot)\} \end{aligned}$$

where the state vector is $x = [x_1, \ldots, x_n]^T \triangleq [SOC_1, \ldots, SOC_n]^T \in \mathbb{R}^n$, the input is $v = [v_1, \ldots, v_n]^T \triangleq [I_{B_1}, \ldots, I_{B_n}]^T \in \mathbb{R}^n$, the output is $y = [y_1, \ldots, y_n]^T \triangleq [V_{B_1}, \ldots, V_{B_n}]^T \in \mathbb{R}^n$, and $diag\{\cdot\}$ denotes the diagonal matrix.

8.1.2 Charging Constraints

In order to ensure the stability of the battery pack system and prolong its service life, three constraints should be met in the charging procedure, namely, the cells' SOC, charging current and terminal voltage limitations.

SOC constraint: The SOCs of the cells must be lower than their upper limitation x_u to avoid overcharging of the battery pack as:

$$\chi = \{x(k) \in \mathbb{R}^n | x(k) \leq x_u\}. \tag{8.5}$$

Charging current limitation: Since high charging currents are harmful for batteries, the cells' charging currents should be kept in a suitable range ν as

$$\nu = \{v(k) \in \mathbb{R}^n | 0_n \leq v(k) \leq v_M\} \tag{8.6}$$

where $v_M \in \mathbb{R}^n$ is the cells' maximum allowed charging current vector, and 0_n represents the vector with n zeros.

Terminal voltage restriction: The terminal voltages of the cells in the battery pack must be lower than its upper limit. According to (8.4), for battery packs, the following constraint should be satisfied that

$$f(x(k)) + h(x(k))v(k) \leq y_M \tag{8.7}$$

where $y_M \triangleq V_{\max} 1_n \in \mathbb{R}^n$ with $V_{\max}$ the maximum allowed terminal voltage of the cells in the battery pack.

8.2 User-Involved Charging Control Design for Battery Packs

The objectives and constraints related to the batter pack charging is generated and then an optimal control strategy will be formulated in this section to make the battery pack charging system satisfy these objectives and constraints. Figure 8.1 shows the block diagram of the designed optimal multi-objective charging current scheduling strategy with considering user demand, cell equalization, temperature rise, and safety-related constraints. Then, the designed multi-module charger (as in Chap. 7) provides the corresponding charging currents for the battery pack.

The battery pack model (8.4) can be discretized by keeping the cells' charging currents constant for each scheduled sampling interval to better schedule the optimal charging currents of the cells in the battery pack. Then, the discrete model of the battery pack can be obtained as follows:

$$\begin{aligned} x(k+1) &= x(k) + Bv(k) \\ y(k) &= f(x(k)) + h(x(k))v(k) \end{aligned} \tag{8.8}$$

where $B = diag\{\frac{\eta_1 T}{Q_1}, \ldots, \frac{\eta_n T}{Q_n}\} \in \mathbb{R}^{n \times n}$, and T is the scheduling sampling period.

8.2.1 Charging Objectives

Charging task that satisfies the user demand: For a battery pack with the initial cells' SOC vector $x(0) = x_0$, the user can define his or her target SOC Γ_{set} and the charging duration T_{set} according to the next need. Then, the cells' desired SOC vector can be formulated as:

$$x_s(N) = \Gamma_{set} 1_n \text{ with } T_{set} = NT \tag{8.9}$$

where 1_n is a column vector with n ones and N represents the sampling step number. To satisfy this charging task, it is intended to minimize the difference between $x(N)$

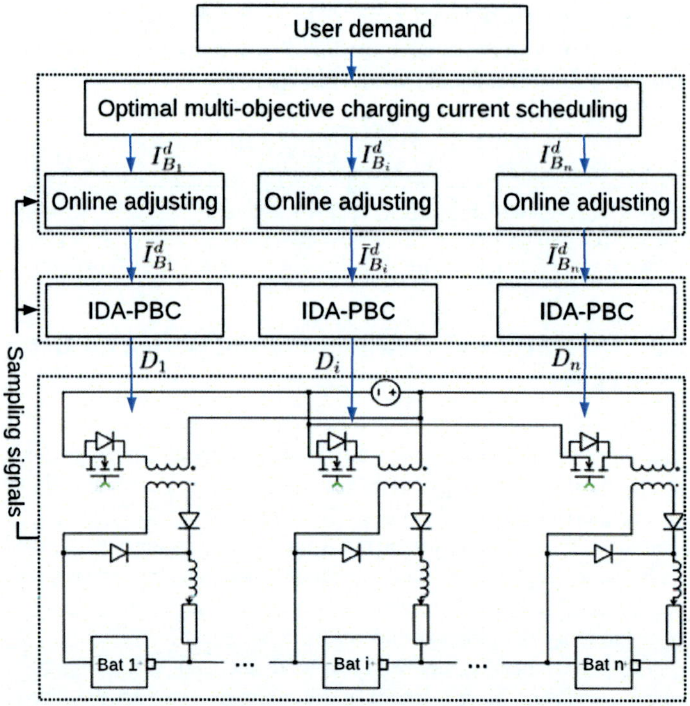

Fig. 8.1 Block diagram of the overall battery charging process

and $x_s(N)$. Hence, the cost function related to the charging task that satisfies the user demand can be formulated as follows:

$$J_x = \frac{1}{2}(x(N) - x_s(N))^T(x(N) - x_s(N)). \tag{8.10}$$

Note that user demands may not be always achieved in practice, for example, the battery pack cannot be fully charged within the very short charging duration required by the user, even when the battery pack is charged with the maximum allowable current. Hence, we choose the cost function (8.10) rather than the hard constraint $x(N) = x_s(N)$ in our charging control strategy, which can also drive $x(N)$ to $x_s(N)$ to the utmost extent while satisfying all the charging constraints.

Cell equalization: Due to the imperfection of battery manufacturing technology and the uneven distribution of temperature, there exist energy imbalance among cells in the battery pack, which results in that the minimum/maximum SOC of the cells limits the available/rechargeable capacity of the entire battery pack. Therefore, in order to improve the effective capacity of the battery pack, it can be achieved by balancing the SOC of each cell to the same level. It is necessary to minimize the cells' SOC difference $||x(k) - \bar{x}(k)||$ during charging, where $|| \cdot ||$ is the 2-norm, and $\bar{x}(k) = \frac{1}{n}1_n 1_n^T x(k)$ denotes the average SOC vector of the cells in the battery pack. Therefore, the cost function related to cell equalization can be obtained as follows:

$$J_e = \frac{1}{2N} \sum_{k=1}^{N} x^T(k) D^T D x(k) \tag{8.11}$$

with

$$D = I_n - \frac{1}{n}1_n 1_n^T$$

where I_n is the identity matrix with the dimension of $n \times n$. It should be pointed out that $\frac{1}{N}$ is included in (8.11) to make the order of magnitude of J_e comparable to J_x in (8.10).

Temperature rise: Referring to [2], the i-th cell's thermal dynamics can be represented as:

$$\begin{aligned} T_{e_i}(k+1) = T_{e_i}(k) - b_{i1}(T_{e_i}(k) - T_a) + b_{i2} v_i^2(k) \\ + \sum_{j=1}^{n} r_{ij}(T_{e_j}(k) - T_{e_i}(k)) \end{aligned} \tag{8.12}$$

where $T_{e_i}(k)$ and T_a represents the i-th cell's lumped temperature and ambient temperature, respectively. r_{ij} denotes the heat transfer coefficient from the j-th cell to the i-th cell in the battery pack. b_{i1} and b_{i2} represent the coefficients relevant to heat transfer and Joule heating, respectively. The cells' temperature rises over a sampling period can be defined as $\Delta T_e(k) = T_e(k+1) - T_e(k)$, where $T_e(k) = [T_{e_1}(k), \ldots, T_{e_n}(k)]^T \in \mathbb{R}^n$. From (8.12), the cells' temperature rises can be calculated as:

$$\Delta T_e(k) = A_e T_e(k) + B_{e1} T_a + B_{e2} \underline{v}(k) \tag{8.13}$$

where

$$A_e = \begin{bmatrix} -b_{11} + r_{11} - \sum_{j=1}^{n} r_{1j} & r_{12} & \cdots & r_{1n} \\ r_{21} & -b_{21} + r_{22} - \sum_{j=1}^{n} r_{2j} & \cdots & r_{2n} \\ \cdots & \cdots & \cdots & \cdots \\ r_{n1} & r_{n2} & \cdots & -b_{n1} + r_{nn} - \sum_{j=1}^{n} r_{nj} \end{bmatrix}$$

$$B_{e1} = [b_{11}, b_{21}, \ldots, b_{n1}]^T$$

$$B_{e2} = diag\{b_{12}, b_{22}, \ldots, b_{n2}\}$$

$$\underline{v}(k) = [v_1^2(k), \ldots, v_n^2(k)]^T.$$

The cells' temperature rises can be restricted by using (8.13) to construct a cost function in the charging control strategy. But the parameters r_{ij}, b_{i1}, and b_{i2} in (8.13) are hard to be accurately identified in practical applications, so it is difficult to obtain the cells' accurate temperature rises in the battery pack. Moreover, utilizing the battery pack's thermal model as in (8.13) will result in the great increase of computational cost of the battery pack charging control algorithm. From (8.13), it shows that the cells' temperature rises are related to their charging currents. Therefore, in order to simplify the problem of temperature rise management, we can limit the magnitude of the square of charging current instead of directly limiting the temperature rise, which leads to the following cost function:

$$J_v = \frac{1}{2N} \sum_{k=0}^{N-1} v^T(k)v(k). \tag{8.14}$$

In general, we do not need to limit the temperature in the charging current design algorithm, because when the battery current and terminal voltage are strictly below the maximum allowable value, the temperature of the battery pack will not exceed the limit. Use (8.14) to ensure that the temperature of the battery pack is not too high. For the sake of thermal safety, the charging process will be terminated when the battery temperature exceeds the limit.

Multi-objective formulation: For the optimal battery pack charging control problem, it is necessary to fully balance user demand, cell equalization, and thermal rise of the battery pack. Then, based on (8.10), (8.11), and (8.14), the following quadratic multi-objective cost function can be obtained as:

$$J(x(k), v(k)) = \gamma_1 J_x + \gamma_2 J_e + \gamma_3 J_v \tag{8.15}$$

where $\gamma_1 \geq 0$, $\gamma_2 \geq 0$, and $\gamma_3 \geq 0$ are the trade-off weights.

8.2.2 Optimal Battery Pack Charging Control Design

With taking the multi-objective (8.15) and the constraints (8.5)–(8.7) into consideration, the battery pack charging control problem can be transformed to the following constrained optimization problem:

$$\begin{aligned} &\underset{v(0),\ldots,v(N-1)}{\text{minimize}} \quad J(x(k), v(k)) \\ &\text{s.t. } x(k+1) = x(k) + Bv(k), \;\; x(0) = x_0 \\ &\qquad f(x(k)) + h(x(k))v(k) \leq y_M \\ &\qquad x(k) \leq x_u, \;\; v(k) \in \nu. \end{aligned} \tag{8.16}$$

According to (8.16), it can be obtained that

$$x(k) = x_0 + \textstyle\sum_{j=0}^{k-1} Bv(j) \tag{8.17}$$

With defining $U \triangleq [v^T(0), \ldots v^T(N-1)]^T \in \mathbb{R}^{nN}$ and $H_k \triangleq [\Upsilon_k, \Theta_k] \in \mathbb{R}^{n \times nN}$ with $\Upsilon_k = [I_n, \ldots, I_n] \in \mathbb{R}^{n \times kn}$ and $\Theta_k = [0_{n \times n}, \ldots, 0_{n \times n}] \in \mathbb{R}^{n \times (N-k)n}$, (8.17) can be represented as

$$x(k) = x_0 + BH_k U.$$

Based on this, the constrained optimization problem (8.16) can be rewritten in the following formulation:

$$\begin{aligned} \underset{U}{\text{minimize}} \ & J_1(U) + \tau \\ \text{s.t.} \ \ & F(U) + G(U)U \leq Y_M \\ & MU \leq X_C, \ \Phi U \leq U_M \end{aligned} \tag{8.18}$$

with the auxiliary variables

$$\begin{aligned} J_1(U) &= x_0^T \Psi_1 U - x_s^T \Psi_2 U + U^T \Psi_3 U \\ \tau &= \frac{\gamma_1}{2}(x_0^T x_0 - 2x_0^T x_s(N) + x_s^T(N) x_s(N)) + \frac{\gamma_2}{2} x_0^T D^T D x_0 \\ \Psi_1 &= \gamma_1 B H_N + \frac{\gamma_2}{N} \sum_{j=1}^{N} D^T D B H_j, \ \Psi_2 = \gamma_1 B H_N \\ \Psi_3 &= \frac{\gamma_1}{2} H_N^T B^T B H_N + \frac{\gamma_2}{2N} \sum_{j=1}^{N} H_j^T B^T D^T D B H_j + \frac{\gamma_3}{2N} I_{nN} \end{aligned}$$

where U is the optimization variable; $Y_M \in \mathbb{R}^{nN}$, $X_C \in \mathbb{R}^{nN}$, $F(U) \in \mathbb{R}^{nN}$, $G(U) \in \mathbb{R}^{nN \times nN}$, $M \in \mathbb{R}^{nN \times nN}$, $\Phi \in \mathbb{R}^{2nN \times nN}$, and $U_M \in \mathbb{R}^{2nN}$ are defined as:

$$\begin{aligned} Y_M &= [y_M^T, \ldots, y_M^T]^T, \ X_C = [(x_u - x_0)^T, \ldots, (x_u - x_0)^T]^T \\ F(U) &= [f^T(x_0 + DH_1 U), \ldots, f^T(x_0 + DH_N U)]^T \\ G(U) &= diag\{h(x_0 + DH_1 U), \ldots, h(x_0 + DH_N U)\} \\ M &= [(BH_1)^T, \ldots (BH_N)^T]^T, \ \Phi = [I_{nN}, -I_{nN}]^T \\ U_M &= [v_M^T, \ldots, v_M^T, 0_n^T, \ldots, 0_n^T]^T. \end{aligned}$$

Because τ is a constant, it can be removed to solve the optimization problem (8.18). The convex barrier function [3] is introduced to replace the inequality constraints. Then, the constrained optimization problem (8.18) can be transformed into:

$$\underset{U}{\text{minimize}} \ J_1(U) - \frac{1}{\mu_1} \sum_{j=1}^{4nN} \log(-g_j(U)) \tag{8.19}$$

where $-\sum_{j=1}^{4nN} \log(-g_j(U))$ denotes the logarithmic barrier function, μ_1 denotes a positive parameter, and $g_j(U)$ denotes the j-th element of the vector $g(U) \in \mathbb{R}^{4nN}$ given by

$$g(U) = \begin{bmatrix} F(U) + G(U)U - Y_M \\ MU - X_C \\ \Phi U - U_M \end{bmatrix}. \tag{8.20}$$

The algorithm based on interior point in [3] is used to solve the optimization problem (8.19), and then the optimal charging current sequence $I_{B_i}^d$ $(1 \leq i \leq n)$ is calculated.

Variable scheduling sampling period design: In practical applications, the success of the charging control method is closely related to the selected scheduling sampling period T. With a large T, the model discretization error of (8.8) is correspondingly large, which may make the actual states of the cells easily exceed their limit when the cells are closed to the fully charged state. However, if the scheduling sampling period is selected too small, the corresponding amount of computation will increase sharply, which is not desirable in practice. In order to balance the charging performance and computational burden, a variable scheduling sampling period is designed as follows:

$$T = \begin{cases} T_1 \ x(k) \leq x_T \\ T_2 \ x(k) > x_T \end{cases} \tag{8.21}$$

where $x_T = \Gamma_T 1_n$ with Γ_T a preset SOC that is closed to the fully-charged state. T_1 and T_2 are scheduling sampling periods, which are designed by taking into account both the actual accuracy requirements of discrete time battery model and the computational burden of the charging current scheduling method. From (8.21), it observes that a large T_1 is selected to reduce the computational amount when the cells' SOCs are less than x_T while the scheduling sampling period is changed to a small value T_2 to avoid large model discretization error when the cells' SOCs increase to be larger than x_T. The detailed method for the designed implementation of the variable scheduling sampling period is described in Algorithm 8.1.

Algorithm 8.1

- (1) Set the user demand Γ_{set}, T_{set}, the intial cells' SOC vector x_0, the preset SOC vector x_T, and the scheduling sampling period $T = T_1$. The step number is $N_1 = \frac{T_{set}}{T_1}$.
- (2) Run the interior point algorithm and get the scheduled cells' charging current vector $v(k)$ $(k = 0, \ldots, N_1 - 1)$ and the updated SOC vector $x(k)$ $(k = 1, \ldots, N_1)$.
- (3) If $\Gamma_{set} 1_n \leq x_T$, stop and output $v(k)$ $(k = 0, \ldots, N_1 - 1)$. Otherwise, increase k from 1 to N_1 to find a k_1 that satisfies $k_1 = \max\{k : x(k) \leq x_T\}$. Record $v(k)$ $(k = 0, \ldots, k_1 - 1)$ and set $x(k_1)$ as the new initial cells' SOC vector.
- (4) Change the scheduling sampling period $T = T_2$, and the step number is $N_2 = \frac{T_{set} - k_1 T_1}{T_2}$. Run the interior point algorithm again and get the designed cells' charging current vector $v(k)$ $(k = k_1, \ldots k_1 + N_2 - 1)$.
- (5) Terminate and output $v(k)$ $(k = 0, \ldots, k_1 - 1, k_1 \ldots k_1 + N_2 - 1)$ by combining the results from Step (3) and Step (4).

On-line adjustment strategy of desired charging currents: Because the model used in our designed charging current scheduling method is not completely consistent with the battery pack's actual dynamics, the terminal voltage V_{B_i} of the i-th $(1 \leq i \leq n)$ cell may exceed it upper limit with the predetermined optimal charging current

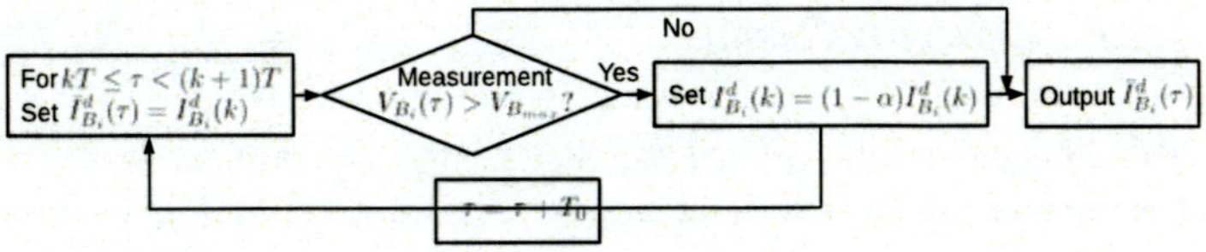

Fig. 8.2 Scheme of the on-line charging current adjustment method

$I^d_{B_i}$. From (8.7), it is observed that the cell's terminal voltage is positively related to its charging current. Therefore, when the measured terminal voltage of the i-th cell is greater than its maximum allowable voltage $V_{B_{\max}}$, the charging current should be reduced to make V_{B_i} decrease. From this view, an on-line charging current adjustment method is proposed, which makes the actual expected charging current decrease at a specific rate of α $(0 < \alpha < 1)$ until the terminal voltage of the i-th cell V_{B_i} satisfies the constraint (8.7) again. Figure 8.2 shows the flow chart of the designed charging current adjustment method, where $\bar{I}^d_{B_i}$ denotes the i-th cell's required charging current after adjustment and T_0 represents the control sampling period. It should be pointed out that when the a large adjustment rate of α is selected, the tuning speed is correspondingly faster and it can cause a shorter duration of the overvoltage. However, it may result in an excessive reduction of the $\bar{I}^d_{B_i}$, which will affect the charging performance of the battery pack. Because the overvoltage phenomenon usually occurs when the cells are closed to be fully charged, the adjusted SOC trajectories of the cells in the battery pack are still close to the predetermined trajectory.

8.3 Simulation Results

Simulations based on MATLAB are performed for a battery pack consisting of 10 serially connected cells to verify the effective of the designed optimal charging current scheduling strategy. The cells' capacities are randomly selected, varying from 1.90 to 2.10 Ah with $[Q_1, \ldots, Q_{10}] = $ [2.07 Ah, 1.91 Ah, 1.93 Ah, 1.96 Ah, 1.97 Ah, 2.09 Ah, 2.06 Ah, 1.99 Ah, 1.95 Ah, 2.10 Ah]. The cells' initial SOCs are randomly set to $[x_1(0), \ldots, x_{10}(0)] = [12\%, 20\%, 18\%, 19\%, 16\%, 14\%, 17\%, 10\%, 15\%, 11\%]$ with the initial SOC difference of $||x(0) - \bar{x}(0)|| = 10.28\%$. The cells' maximum allowed charging current is chosen as 3 C-rate. The cells' OCVs and the internal resistances are selected as in [1]. The ambient temperature is set as 25 °C and the temperature influence coefficients b_{i1} and b_{i2} $(1 \leq i \leq 10)$ are chosen as the same in [2]. The user's desired SOC of the battery pack Γ_{set} and charging duration T_{set} are set to 100% and 60 min/120 min/180 min. The weights in the cost function (8.15) in our designed battery pack charging control algorithm are chosen as $\gamma_1 = 1000$, $\gamma_2 = 5$, and $\gamma_3 = 0.01$, respectively.

8.3.1 Charging Results

The simulation results of the cell's SOCs, charging currents, terminal voltages, and temperatures of the cells in the battery pack are shown in Fig. 8.3. From Figs. 8.3b, 8.4b, and 8.5b, it observes that the optimal charging current scheduled by our designed charging method can be adjusted according to different user settings. For a tight charging demand of 60 min, large charging currents are designed to charge the battery pack to the fully charged state to the greatest extent, while small charging currents are scheduled for a sufficient charging demand of 180 min. Figures 8.3c, 8.4c, and 8.5c illustrates the terminal voltage responses of the cells, from which it can be observed that the constraint (8.7) can be well satisfied when using the designed charging current. Table 8.1 illustrates the average SOC and SOC difference of the cells in the battery pack at the end of the charging process, as well as the maximum temperature

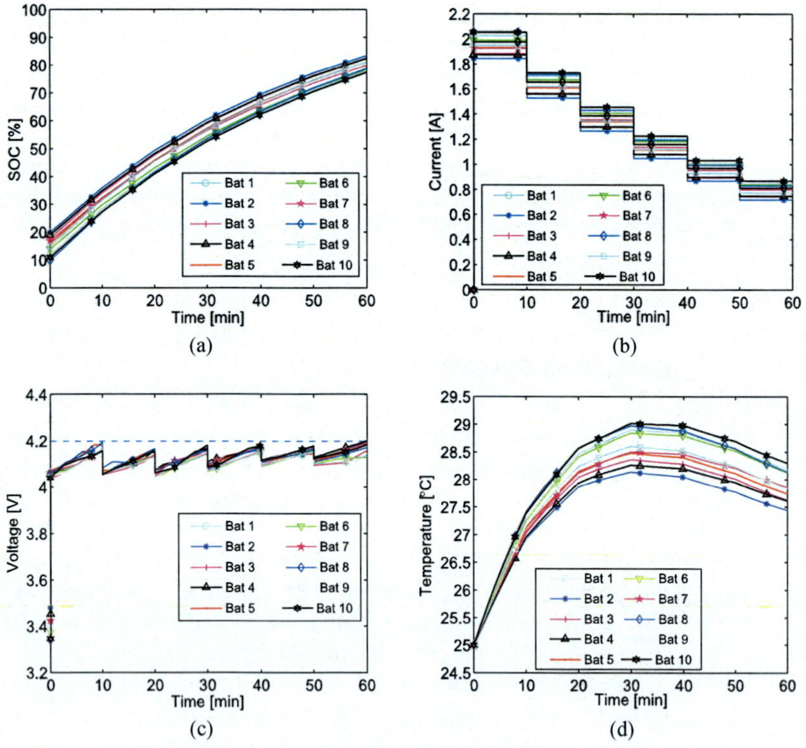

Fig. 8.3 Simulation results of cells' **a** SOCs, **b** charging currents, **c** terminal voltages, **d** temperatures with $T_{set} = 60$ min

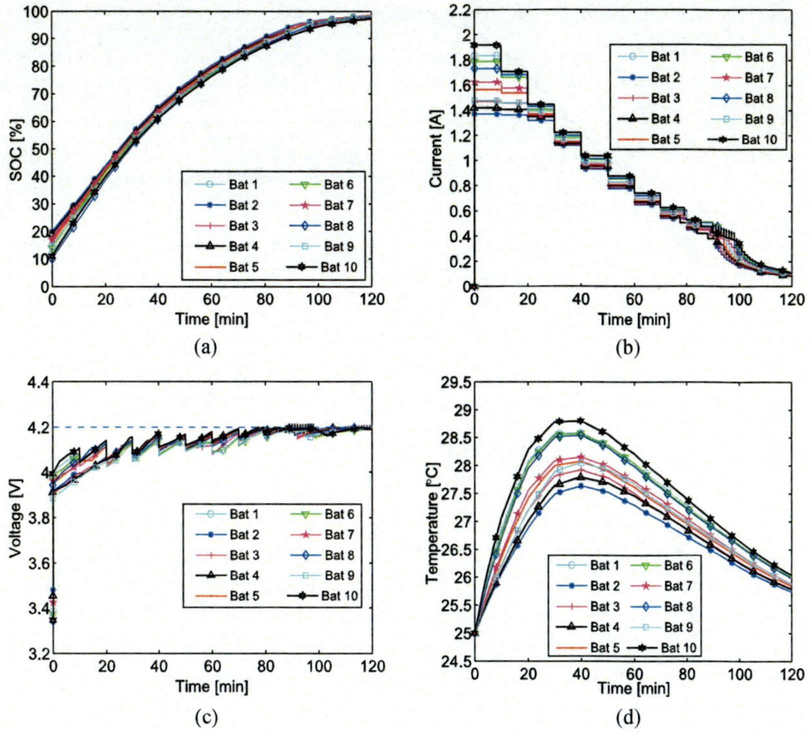

Fig. 8.4 Simulation results of cells' **a** SOCs, **b** charging currents, **c** terminal voltages, **d** temperatures with $T_{set} = 120$ min

Table 8.1 Simulation results for different user settings

	Average SOC (%)	SOC difference (%)	Maximum temperature (°C)
Initial	15.2	10.83	–
60 min charging	80.5	6.67	29.01
120 min charging	98.01	1.85	28.8
180 min charging	99.36	1.28	26.81

during the charging process for different user settings. It shows that the cells' average SOC can be charged to close to the desired value with the scheduled optimal charging currents. The effectiveness of our proposed optimal charging control method on cell equalization is also demonstrated.

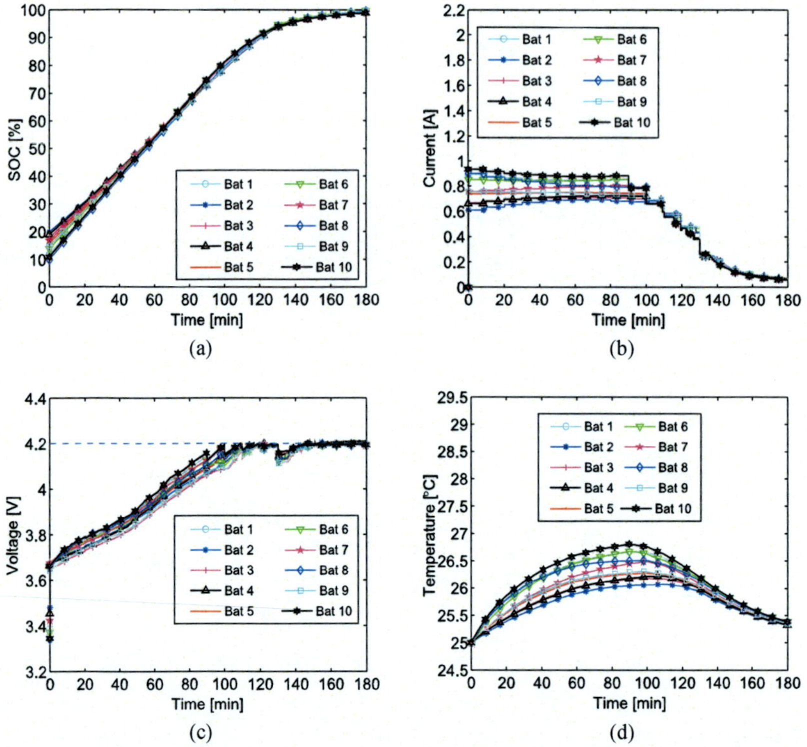

Fig. 8.5 Simulation results of cells' **a** SOCs, **b** charging currents, **c** terminal voltages, **d** temperatures with $T_{set} = 180$ min

8.3.2 High Current Charging

Because of the large internal resistance of the cells, the designed charging currents are limited less than 1 C-rate to avoid overvoltage of the cells as shown in Figs. 8.3, 8.4 and 8.5, although the maximum allowable charging current is set to 3 C-rate. In the simulation, the internal resistance is set to 10% of the original value to verify the performance of the designed charging scheme at a higher C-rate. We set the required SOC to $\Gamma_{set} = 100\%$ and set the charging time to $T_{set} = 20$ min. Figure 8.6a–d show the simulation results in terms of cell's SOCs, charging currents, terminal voltages, and temperature, respectively. The maximum charging current of the cells can be as high as 3 C-rate. The cells' average SOC increases from 10 to 94.14% after $T_{set} = 20$ min charging. The maximum temperature that the cells reaches about 38.8°C, which meets the actual thermal constraints. The results show that fast charging with all the charging constraints satisfied can also be achieved by setting a short charging duration in our designed charging control method.

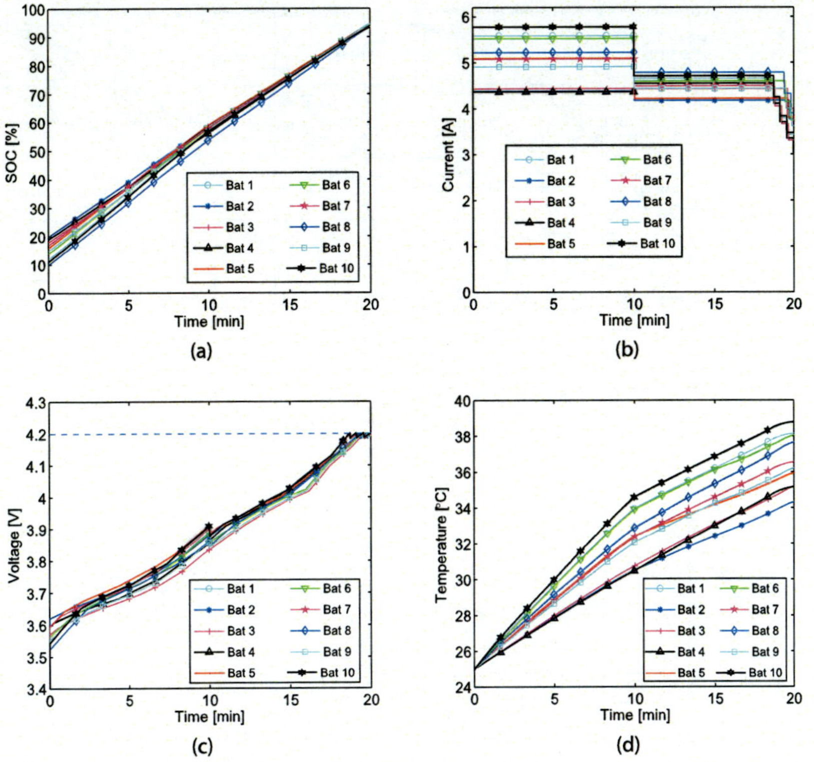

Fig. 8.6 Simulation results of cells' **a** SOCs with $T_{set} = 20$ min, **b** charging currents with $T_{set} = 20$ min, **c** terminal voltages with $T_{set} = 20$ min, **d** temperatures with $T_{set} = 20$ min

8.3.3 Effect Analysis of Weight Selection

In the multi-objective cost function (8.15), the weight coefficient γ_i $(1 \leq i \leq 3)$ represents the relative importance of each goal, where larger γ_1, γ_2, and γ_3 place more emphasis on the consideration of charging task that satisfies the user demand, cell balance, and temperature rise, respectively. In the actual charging procedure, the charging task that satisfies the user demand is the most important, so a large γ_1 is needed. Then, different γ_2 and γ_3 are selected to verify the weight selection on the charging performance.

First, γ_1 and γ_3 are set to 1000 and 0.01, and γ_2 is chosen as 0.05, 0.1, 0.5, 1, 5, 10, 20, and 30, respectively. Figure 8.7a, c, and e show the simulation results of the average SOC, SOC difference, and average temperature of the battery pack, respectively. It is observed that a less SOC difference can be obtained with the selection of a large γ_2, which is consistent with the analysis. However, when γ_2 is selected too large, it may also degrade the performance of the charging task that

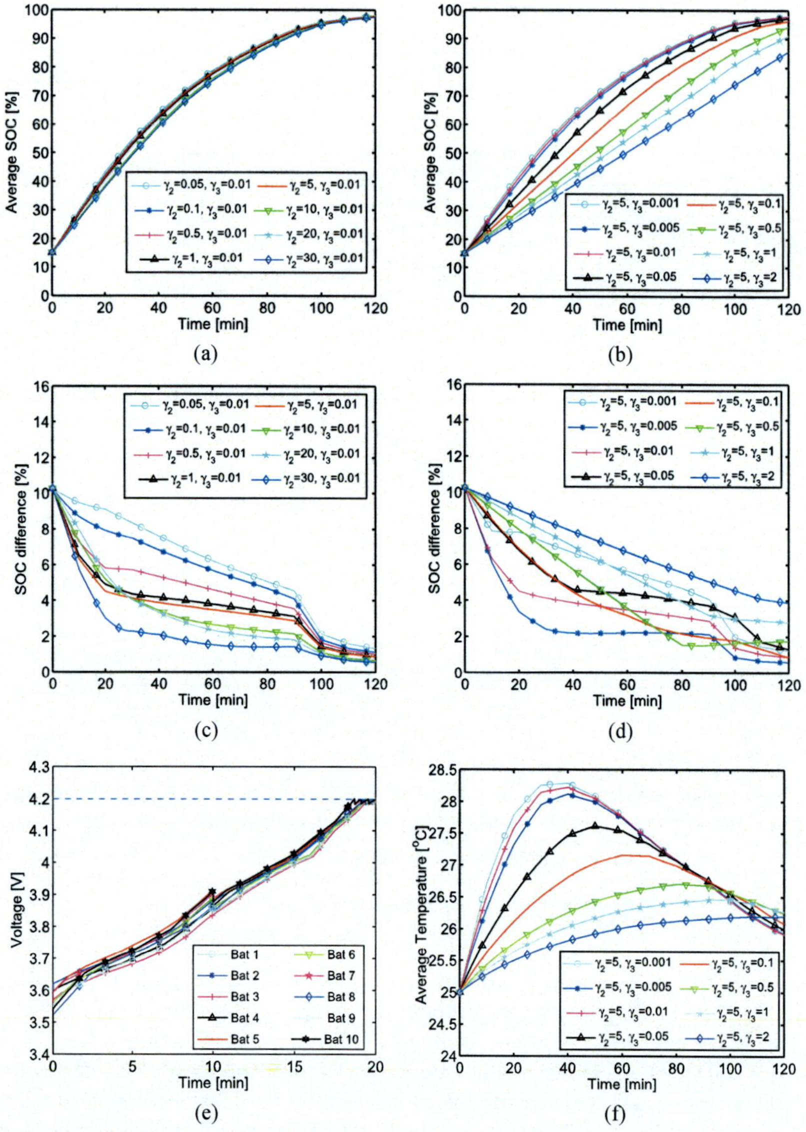

Fig. 8.7 Simulation results of cells' **a** average SOCs with different γ_2, **b** average SOCs with different γ_3, **c** SOC differences with different γ_2, **d** SOC differences with different γ_3, **e** average temperatures with different γ_2, **f** average temperatures with different γ_3

satisfies the user demand. Next, γ_1 and γ_2 are chosen as 1000 and 5, and a new set of γ_3 is set to 0.001, 0.005, 0.05, 0.1, 0.5, 1, and 2, respectively. The results are illustrated in Fig. 8.7b, d, and f, which indicates that a large γ_3 brings a lower average temperature, but may case the battery pack to be insufficiently charged. Overall, it can be concluded that larger γ_2 and γ_3 can bring smaller SOC difference and lower average temperature, respectively, but both may adversely affect the performance of the charging task that satisfies the user demand. In practical applications, users can take this as the guide to select appropriate weights to balance these goals.

8.4 Experimental Results

Figure 8.8 shows the constructed experimental bench, which includes a battery pack composed of 3 PowerFocus 18650 lithium-ion batteries connected in series, a set of dSPACE, 3 OMEGA temperature sensors, and an NI general purpose inverter controller (GPIC) Single-Board 9683. The cells' capacities are $Q_1 = 2.341$ Ah, $Q_2 = 2.387$ Ah, and $Q_3 = 2.379$ Ah, respectively. Figure 8.9a shows the relationship between OCVs and SOCs of the cells and Fig. 8.9b illustrates the cells' internal resistances. When the cells' SOCs are within $[5\%, 95\%]$, their internal resistances are approximately constant, but when their SOCs are close to the fully charged state of $[95\%, 100\%]$, their internal resistances increase sharply, indicating that the terminal voltages of the cells can easily exceed their upper bound when their SOCs are greater than 95%. Hence, the preset SOC and the sampling periods are, respectively, selected as $\Gamma_T = 95\%$, $T_1 = 600$ s, and $T_2 = 60$ s in the variable scheduling sampling method. The upper bounds of the cells' SOCs and terminal voltages are $x_u = 100\%$ and $V_{\max} = 4.2$ V, respectively. When the terminal voltage of the cell is greater than 4.2 V, the scheduled charging current is reduced by 5% at each step until the its terminal voltage goes below the limiting condition (8.7), where the sampling period is set as $T_0 = 0.5$ s. The modified isolated buck converter based multi-module battery pack charger as mentioned in Chap. 7 is modeled in the dSPACE through real-time simulation.

Several experiments are carried out on the above-mentioned three-modular battery pack to verify the performance of the designed optimal multi-objective charging control method. The cells' initial SOCs are set to $[5\%, 15\%, 10\%]$. The maximum allowable charging current of the cells is 0.5 C-rate recommended by the manufacturer. The demanded battery pack's SOC and charging duration are selected as $\Gamma_{set} = 100\%$ and $T_{set} = 210$ min. The proposed charging current scheduling algorithm runs on the GPIC board.

Because the designed optimal charging current scheduling algorithm uses a simple Rint battery model, the model bias may make the terminal voltages of the cells easily exceed their limitation (8.7) with the scheduled charging currents if the cells are nearly fully charged. As shown in Figs. 8.10a and c, the terminal voltages of the cells can be quickly pushed to below the preset limitation of 4.2 V through reducing the charging currents by a specific rate $\alpha = 5\%$ each time. The results

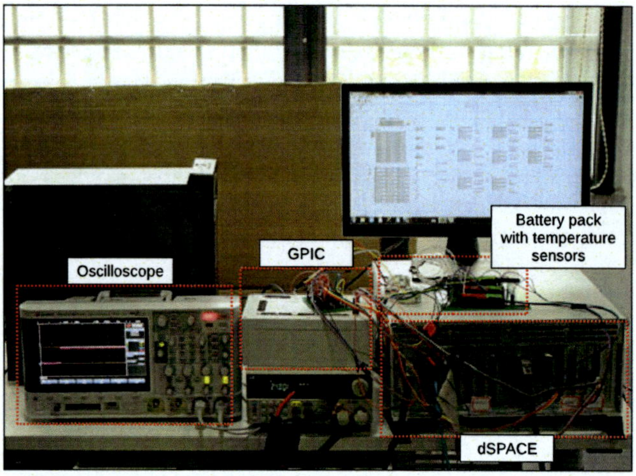

Fig. 8.8 Experimental test bench of the battery pack charging system

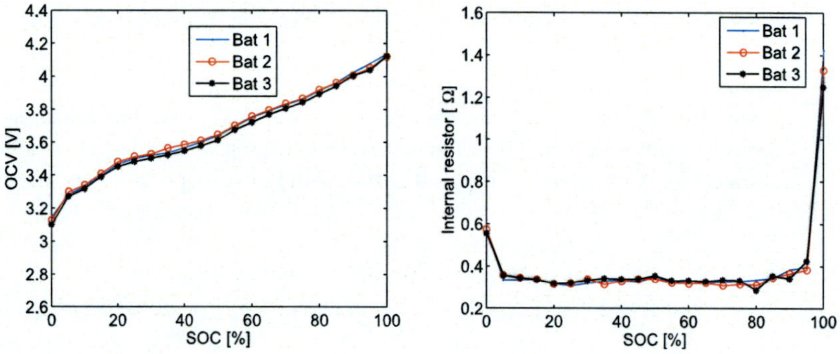

(a) (b)

Fig. 8.9 **a** Relationship between the cells' OCVs and SOCs, **b** mapping from the SOCs to the internal resistors of the cells

show the superiority of the designed on-line charging current adjustment method. Figure 8.10b shows the expected and actual charging current responses of the multi-module charger, which denotes that the charger can provide the required charging current well. Figure 8.4 shows the SOC responses of the cells in the battery pack. At the end of the charging procedure, the average SOC of the battery pack is 99.16%, where $SOC_1 = 99.45\%$, $SOC_2 = 98.94\%$, and $SOC_3 = 99.09\%$. It is very close to the expected value (100%). The cells' SOC difference is reduced from 7.07% to 0.37%. The results show that the proposed optimal multi-objective charging strategy can meet the needs of users and the charging goal of cell equalization.

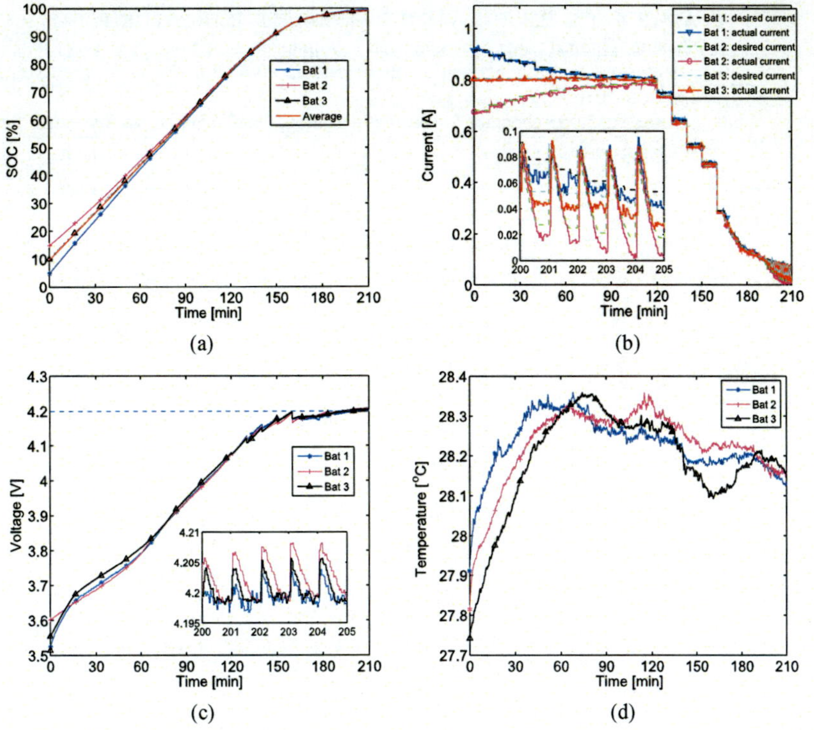

Fig. 8.10 Experimental results of **a** cells' SOC responses, **b** charging currents, **c** terminal voltages, **d** temperatures

If traditional chargers are utilized, the charging currents of all cells in the battery pack are the same. The charging procedure must end when the second cell is fully charged ($SOC_2 = 100\%$) to avoid its overcharging, but the SOCs of the first and third cells are $SOC_1 = 91.67\%$ and $SOC_3 = 95.29\%$, respectively. The average SOC of the battery pack is 95.65% and SOC difference is 5.91%. Compared with the traditional charger, the charging strategy with cell equalization proposed here can effectively improve the effective capacity of the battery pack. As shown in Fig. 8.10d, the temperatures of the cells measured by the OMEGA temperature sensor can be kept within an appropriate range of [27.7 °C, 28.4 °C] during charging.

References

1. Q. Ouyang, J. Chen, J. Zheng, and H. Fang, "Optimal multiobjective charging for lithium-ion battery packs: A hierarchical control approach," *IEEE Transactions on Industrial Informatics*, vol. 14, no. 9, pp. 4243–4253, 2018.

2. A. Abdollahi, X. Han, G. Avvari, N. Raghunathan, B. Balasingam, K. Pattipati, and Y. Bar-Shalom, "Optimal battery charging, part I: Minimizing time-to-charge, energy loss, and temperature rise for OCV-resistance battery model," *Journal of Power Sources*, vol. 303, pp. 388–398, 2016.
3. S. Boyd, L. Vandenberghe, *Convex Optimization*, Cambridge University Press, New York, NY, USA, 2004.

Chapter 9
User-Involved Charging Control for Battery Packs: Leader-Followers Structure

In Chap. 8, an attempt was made to propose an optimal charging control method for battery packs with a centralized structure, which is based on a multi-module charger that allows each cell in the pack to be charged independently. Although good charging results have been achieved, the centralized structure can bring a significant increase in the computational effort of the charging algorithm since it needs to schedule one optimal charging current for each cell in the battery pack. In addition, the cell's model bias cannot be well handled by this method, since it is an open-loop structure.

Considering such a gap, we intend to propose an optimal charging control method for battery packs with a reasonable computational complexity in this Chapter. With utilizing a multi-module charger, a user-involved method with leader-followers framework is designed for the charging control of a battery pack [1]. Specifically, a nominal model of the cells is first utilized to formulate and solve a multi-objective optimization problem to schedule an optimal average SOC trajectory with full consideration of user requirements and energy loss of the battery pack. Next, a distributed charging method with observers for on-line compensation of the cells' model bias is designed to enable the cells' SOCs (followers) track the pre-scheduled average SOC trajectory (leader).

9.1 Charging Model and Constraints

For the battery pack with n cells connected in series, a multi-module charger is developed as shown in the Fig. 7.3, in which each cell can be charged by a small charging module in parallel. It can essentially ensure the adequate charging of the battery pack without any risk of overcharging. In this study, we focus on the optimal charging current design and ignore the dynamics of the battery pack charger.

Q. Ouyang and J. Chen, *Advanced Model-Based Charging Control for Lithium-Ion Batteries*, https://doi.org/10.1007/978-981-19-7059-7_9

9.1.1 Battery Pack Model

For a battery pack with n cells connected in series, the Rint model that characterizes the dynamics of each cell is adopted in the charging control algorithm design, as shown in Figs. 3.5 and 7.3. It is worth noting that the designed charging control method can be easily extended to any other cell's models. Figure 7.3 shows the i-th ($1 \leq i \leq n$) cell's model, which consists of two parts, i.e., a voltage source that simulates energy storage and a serially connected internal resistance that represents the energy loss in charging/discharging. The i-th cell's model is

$$\begin{aligned} SOC_i(k+1) &= SOC_i(k) + b_i I_{B_i}(k) \\ V_{B_i}(k) &= V_{OC_i}(k) + R_{0_i}(k) I_{B_i}(k) \end{aligned} \tag{9.1}$$

where $b_i = \frac{\eta_0 T}{Q_i}$; $SOC_i(k)$, $I_{B_i}(k)$, $V_{B_i}(k)$, and Q_i denote the i-th cell's SOC, charging current, terminal voltage, and capacity respectively; η_0 and T are the Coulombic efficiency and sampling period, respectively; $V_{OC_i}(k)$, and $R_{0_i}(k)$ represent the OCV and the internal resistance of the i-th cell, which can be expressed as follows:

$$V_{OC_i}(k) = f_i(SOC_i(k)), \quad R_{0_i}(k) = h_i(SOC_i(k)). \tag{9.2}$$

Based on (9.1)–(9.2), the model of the battery pack consisting of n serially connected cells can be represented as:

$$\begin{aligned} x(k+1) &= x(k) + Bu(k) \\ y(k) &= f(x(k)) + h(x(k))u(k) \end{aligned} \tag{9.3}$$

where the input, state, and output vectors are $u(k) = [u_1(k), \ldots, u_n(k)]^T \triangleq [I_{B_1}(k), \ldots, I_{B_n}(k)]^T \in \mathbb{R}^n$, $x(k) = [x_1(k), \ldots, x_n(k)]^T \triangleq [SOC_1(k), \ldots, SOC_n(k)]^T \in \mathbb{R}^n$, and $y(k) = [y_1(k), \ldots, y_n(k)]^T \triangleq [V_{B_1}(k), \ldots, V_{B_n}(k)]^T \in \mathbb{R}^n$, respectively; $f(\cdot) = [f_1(\cdot), \ldots, f_n(\cdot)]^T \in \mathbb{R}^n$, $B = diag\{b_1, \ldots, b_n\} \in \mathbb{R}^{n \times n}$, and $h(\cdot) = diag\{h_1(\cdot), \ldots, h_n(\cdot)\} \in \mathbb{R}^{n \times n}$ with $diag\{\cdot\}$ representing the diagonal matrix.

For a battery pack, the energy among cells is normally unbalanced, which results in the battery pack's available capacity being determined by the cell with the lowest SOC. It performs a behavior of buckets effect, which can be expressed as:

$$x_p(k) = min\{x_i(k), \text{for } i = 1, 2, \ldots n\} \tag{9.4}$$

where $x_p(k) \in \mathbb{R}$ represents the SOC of the entire battery pack. The cells' SOC difference, denoted as $x_d(k) \in \mathbb{R}$, can be expressed as follows:

$$x_d(k) = \|x(k) - \bar{x}(k)\| \tag{9.5}$$

where $\|\cdot\|$ denotes the 2-norm, $\bar{x}(k) = \frac{1}{n} 1_n 1_n^T x(k)$ is the average vector of the cell's SOC and 1_n denotes the column vector with n ones.

9.1.2 Safety-Related Charging Constraints

During the battery pack charging procedure, careful consideration of hard constraints such as charging current, SOC, and terminal voltage of each cell in the pack is required to guaranteed the safety of the battery pack.

Charging current limitations: If the charging current is too high, it will affect the performance of the battery and even cause a fire, so the threshold of the charging current is very important, which is of a great importance in the safety of the battery. In view of this, the charging currents of the cells must be kept within a reasonable range as follows:

$$0_n \leq u(k) \leq u_M 1_n \tag{9.6}$$

where $u_M \in \mathbb{R}$ stands for the maximum allowed charging current of the cells in the battery pack and 0_n denotes the column vector with n zeros.

SOC restrictions: In order to avoid overcharging, the SOCs of the cells in the battery pack cannot exceed their upper limits. Hence, the following constraints should be guaranteed that

$$x(k+1) \leq x_M 1_n \tag{9.7}$$

where $x_M \in \mathbb{R}$ represents the upper limits of the cells' SOCs in the battery pack.

Terminal voltage constraints: To avoid damage, the terminal voltage of the battery at the end of each sampling interval at the designed charging current should be less than the allowed voltage limit, based on (9.3), which should satisfy

$$f(x(k+1)) + h(x(k+1))u(k) \leq y_M 1_n \tag{9.8}$$

where $y_M \in \mathbb{R}$ represents the cells' maximum allowed terminal voltage.

9.2 User-Involved Optimal Charging Control Design

After giving the model and hard constraints on battery pack charging, this section first presents the charging task with user participation, and then describes the designed optimal battery pack charging control method in detail.

9.2.1 User-Involved Charging Task Formulation

Most of the existing charging strategies in practice focus on fast charging using high C-rate current to charge the battery. However, high charging current can accelerate the degradation of the battery's capacity. If the user's needs are incorporated into the design of the charging control algorithm, the charging current can be self-adjusted according to the user specification, which will make the charging procedure more intelligent while ensure the safety of the battery pack. In view of this, the user can set his or her demanded battery pack's SOC x_s and charging time T_s in the charging protocol based on his or her future requirements, which can be expressed as

$$x(N) = x_s 1_n \text{ with } T_s = NT \tag{9.9}$$

where N denotes the sampling step number in the charging control algorithm.

Since the hard constraints (9.6)–(9.8) needs to be satisfied throughout the charging procedure, the user's demand is not always satisfied in practice, for example, the battery pack cannot be fully charged from the empty state even if the current used for charging is the maximum allowed. Hence, we do not adopt the hard constraint (9.9) in the battery pack charging control design, but maximize the difference between $x(N)$ and the desired $x_s 1_n$ while ensuring that all constraints are satisfied. Thus, the cost function J_u in terms of the user demand can be expressed as

$$J_u = (x(N) - x_s 1_n)^T (x(N) - x_s 1_n). \tag{9.10}$$

Another important charging goal is to reduce the energy loss during charging procedure [2] to improve the charging efficiency. From (9.3), a new cost function J_e related to the energy loss during charging of the battery pack is generated as follows:

$$J_e = \sum_{k=0}^{N-1} u^T(k) h(x(k)) u(k). \tag{9.11}$$

Therefore, according to (9.3), (9.6)–(9.11) as well as considering both user demand and energy loss, the user-involved charging task can be formulated as a constrained multi-objective optimization issue as follows:

$$\begin{aligned} \min_{u(0),\ldots,u(N-1)} \quad & \gamma_1 J_u + \gamma_2 J_e \\ \text{s.t.} \quad & x(k+1) = x(k) + Bu(k), \quad x(0) \\ & f(x(k+1)) + h(x(k+1))u(k) \le y_M 1_n \\ & 0_n \le u(k) \le u_M 1_n, \ x(k+1) \le x_M 1_n \end{aligned} \tag{9.12}$$

where $x(0)$ represents the initial SOC vector of the cells in the battery pack, γ_1 and γ_2 are the positive weight coefficients. The optimal charging current $u(k)$ $(0 \le k \le N-1)$ with considering the user involvement task for each cell can be calculated by directly solving (9.12). But it can bring the following challenges:

(1) In (9.12), the number of optimization variables is nN, which can result in a great computational burden on the charging controller. It means that the direct calculation method is very difficult to be implemented in practice, especially for the battery pack containing a large number of cells.

(2) The parameters in the battery pack's model (9.3) are often hard to be accurately extracted, and usually only an approximate nominal model can be obtained in practical applications. The direct calculation method based on the model is an open-loop-based strategy that has no ability to compensate for the battery model bias, which may lead to the phenomenon that the designed charging current brings the violation of safety-related constraints.

In order to overcome the above drawbacks, a two-layer optimal charging control method with a leader-follower framework is designed here, where its schematic is shown in Fig. 9.1. An optimal average charging trajectory is calculated by formulating an optimal problem based on the cells' nominal model, and then, a distributed charging control method with on-line compensation of model bias is designed in order to regulate the each cell's charging current to drive all cells' SOC (followers) to track this generated trajectory (leader).

9.2.2 *Optimal Average Charging Trajectory Generation*

Motivated by the leader-followers concept for multi-agent systems [3], we design a distributed average tracking framework for the battery pack's charging, where an average charging trajectory is first designed as the leader, and then all the cells as followers will track this trajectory in parallel. The charging strategy based on this framework has the many advantages over solving (9.12) directly, such as a significant reduction in computational burden. In light of this, the cells' nominal model with their average initial SOC is adopted to schedule the optimal average charging trajectory as:

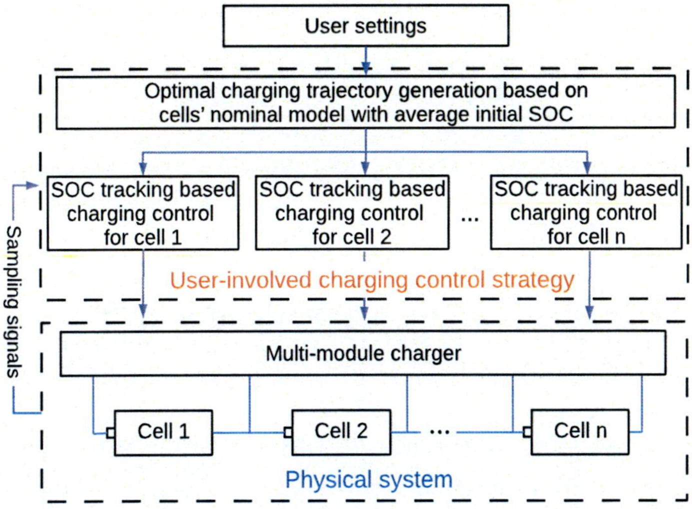

Fig. 9.1 Schematic for user-involved optimal charging control strategy

$$\begin{aligned} x_0(k+1) &= x_0(k) + b_0 u_0(k) \\ y_0(k) &= f_0(x_0(k)) + h_0(x_0(k)) u_0(k) \end{aligned} \tag{9.13}$$

where $x_0(k) \in \mathbb{R}$, $y_0(k) \in \mathbb{R}$, and $u_0(k) \in \mathbb{R}$ represent the state, output, and input of the cells' nominal model in the battery pack, respectively; b_0, $f_0(\cdot)$, and $h_0(\cdot)$ are the nominal values of b_i, $f_i(\cdot)$, and $h_i(\cdot)$ $(1 \le i \le n)$, respectively. The initial value of $x_0(k)$ is defined to the cells' average initial SOC, i.e., $x_0(0) = \frac{1}{n} 1_n^T x(0)$. It should be pointed out that the nominal model of the cells can be a simplified and approximate form of the actual model of the cells. Similar to (9.12),with utilizing the nominal model (9.13), the optimal average SOC trajectory based on user participation can be obtained by solving the following optimization problem:

$$\begin{aligned} \min_{u_0(0),\ldots,u_0(N-1)} \quad & \gamma_1 (x_0(N) - x_s)^2 + \gamma_2 \sum_{k=0}^{N-1} h_0(x_0(k)) u_0^2(k) \\ \text{s.t.} \quad & x_0(k+1) = x_0(k) + b_0 u_0(k) \\ & f_0(x_0(k+1)) + h_0(x_0(k+1)) u_0(k) \le y_M \\ & 0 \le u_0(k) \le u_M, \; x_0(k+1) \le x_M \\ & x_0(0) = \tfrac{1}{n} 1_n^T x(0). \end{aligned} \tag{9.14}$$

With defining $U = [u_0(0), \ldots, u_0(N-1)]^T \in \mathbb{R}^N$ and $H_k = [1_k{}^T, 0_{N-k}{}^T] \in \mathbb{R}^{1 \times N}$, a new formulation of (9.13) can be deduced as

$$x_0(k) = x_0(0) + b_0 H_k U. \tag{9.15}$$

Through substituting $x_0(k)$ with $x_0(0) + b_0 H_k U$, the optimization issue (9.14) can be rewritten as:

$$\begin{aligned} \min_{U} \quad & J_1(U) \\ \text{s.t.} \quad & C(U) \le 0 \end{aligned} \tag{9.16}$$

with

$$\begin{aligned} J_1(U) = U^T (\gamma_1 b_0^2 H_N{}^T H_N + \gamma_2 G(U)) U + 2\gamma_1 b_0 (x_0(0) \\ - x_s) H_N U + \gamma_1 (x_0(0) - x_s)^2 \end{aligned}$$

$$C(U) = \begin{bmatrix} F(U) + G_1(U) U - Y_M \\ \Phi U - U_M \\ M U - X_C \end{bmatrix}$$

where the vector U is the optimization variable; $G(U) \in \mathbb{R}^{N \times N}$, $F(U) \in \mathbb{R}^N$, $G_1(U) \in \mathbb{R}^{N \times N}$, $Y_M \in \mathbb{R}^N$, $\Phi \in \mathbb{R}^{2N}$, $U_M \in \mathbb{R}^{2N}$, $M \in \mathbb{R}^{N \times N}$, and $X_C \in \mathbb{R}^N$, respectively, are

$$G(U) = diag\{h_0(x_0(0)), \ldots, h_0(x_0(0) + b_0 H_{N-1} U)\}$$
$$F(U) = [f_0(x_0(0) + b_0 H_1 U), \ldots, f_0(x_0(0) + b_0 H_N U)]^T$$
$$G_1(U) = diag\{h_0(x_0(0) + b_0 H_1 U), \ldots, h_0(x_0(0) + b_0 H_N U)\}$$
$$Y_M = y_M 1_N, \quad \Phi = [I_N, -I_N]^T, \quad U_M = [u_M {1_N}^T, {0_N}^T]^T$$

where I_N represents the identity matrix with dimensions of $N \times N$. Here, the barrier method [4] is adopted to calculate the optimal charging current sequence U^r. The detailed description of the barrier method is shown in Algorithm 9.1. Then, the optimal average SOC trajectory $x_0^r(k)$ for all $1 \leq k \leq N$ can be scheduled as

$$x_0^r(k) = x_0(0) + b_0 H_k U^r. \tag{9.17}$$

It should be pointed out that the dimension of the optimization variables in the optimal average charging trajectory generation algorithm (9.14) is only N, which is $\frac{1}{n}$ of that in (9.12) for a battery pack consisting n series-connected cells. It means that the designed algorithm can significantly reduce the computational effort, showing the superiority of the proposed leader-followers framework for battery pack charging.

Algorithm 9.1:

- (1) Set the iteration index $t = 0$, the initial optimization variable $U^{(0)}$, the intermediate variable $U = U^{(0)}$, the initial parameter $\mu_1^{(0)} > 0$, the growth value $c > 1$, and the tolerances $\varepsilon_1 > 0$ and $\varepsilon_2 > 0$.
- (2) Let $L = J_1(U) - \frac{1}{\mu_1} \sum_{j=1}^{4N} log(-C_j(U))$, where $C_j(U)$ is the j-th element of the vector $C(U)$. Calculate the Newton step and decrement: $\Delta U \triangleq -(\nabla^2 L)^{-1} \nabla L$ and $\alpha \triangleq (\nabla L)^T (\nabla^2 L)^{-1} \nabla L$ with $\nabla L = \frac{\partial L}{\partial U}(U)$ and $\nabla^2 L = \frac{\partial^2 L}{\partial U^2}(U)$.
- (3) Update $U = U + \zeta \Delta U$ with the step size ζ selected by backtracking line search [4].
- (4) Set $U^{(t)} = U$ if $\frac{\alpha}{2} \leq \varepsilon_1$. Otherwise, return to Step 2).
- (5) Stop and output $U^r = U^{(t)}$ if $4N/\mu_1^{(t)} < \varepsilon_2$. Otherwise, set $t = t + 1$, $\mu_1^{(t+1)} = c\mu_1^{(t)}$ and return to Step 2).

9.2.3 Distributed SOC Tracking-Based Charging Control

The idea of the SOC tracking-based charging control strategy is to use the distributed framework that individually controls the i-th ($1 \leq i \leq n$) cell's current to drive its SOC $x_i(k)$ to follow the optimal trajectory $x_r^0(k)$ pre-generated above. Note that the accurate model parameters extraction of all the cells in the battery pack is very cumbersome, as we need to conduct complicated prior experiments. In addition, in general, factors such as aging can cause a change in the model parameters, which is

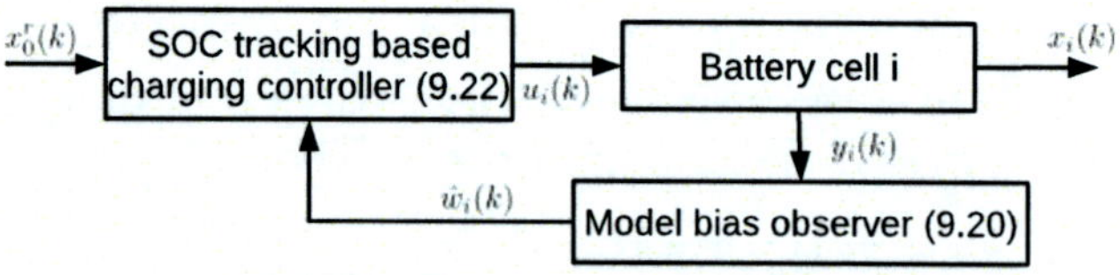

Fig. 9.2 Block diagram of SOC tracking-based charging control strategy for the i-th cell

difficult to be captured accurately. This deficiency can be remedied through adopting the simplified nominal model of the battery in the designed SOC tracking-based charging control method and developing an observer for on-line model bias compensation of each cell. In this view, the model of the i-th cell can be represented as follows:

$$\begin{aligned} x_i(k+1) &= x_i(k) + b_i u_i(k) \\ y_i(k) &= f_0(x_i(k)) + h_0(x_i(k))u_i(k) + w_i(k) \end{aligned} \tag{9.18}$$

with $w_i(k)$ the unknown model bias term that can be expressed as

$$w_i(k) = f_i(x_i(k)) - f_0(x_i(k)) + h_i(x_i(k))u_i(k) - h_0(x_i(k))u_i(k). \tag{9.19}$$

For the i-th $(1 \leq i \leq n)$ cell, we propose a battery pack charging control method based on SOC tracking, as shown in Fig. 9.2, where the deviation of the battery model is estimated by the following designed observer.

Model bias observer: From (9.18), an observer is designed to estimate the i-th cell's unknown model bias as follows:

$$\hat{w}_i(k+1) = \hat{w}_i(k) + l_i(y_i(k) - \hat{y}_i(k)) \tag{9.20}$$

with

$$\hat{y}_i(k) = f_0(x_i(k)) + h_0(x_i(k))u_i(k) + \hat{w}_i(k)$$

where $\hat{w}_i(k)$ denotes the estimation of $w_i(k)$, and l_i represents the designed gain in the observer.

Convergence Analysis: It can be seen from (9.1) that the i-th cell has a slow dynamics, which is due to the fact that in practice the value of Q_i is much larger than the value of $I_{B_i}(k)$. Therefore, the unknown model bias $w_i(k)$ can be assumed to be slowly changing and satisfy $w_i(k+1) \approx w_i(k)$. With defining the estimation error of the model bias as $\tilde{w}_i(k) = w_i(k) - \hat{w}_i(k)$, we can obtain $\tilde{w}_i(k+1) = (1 - l_i)\tilde{w}_i(k)$.

The Lyapunov candidate function is chosen as $V_i(k) = \frac{1}{2}\tilde{w}_i^2(k)$, and its change, denoted as $\Delta V_i(k) = V_i(k+1) - V_i(k)$, can be computed as follows:

$$\Delta V(k) = -(l_i - \tfrac{l_i^2}{2})\tilde{w}_i^2(k). \tag{9.21}$$

From (9.21), it can be concluded that if $0 < l_i < 2$ is selected, $\Delta V(k) \leq 0$ and $\Delta V(k) = 0$ if and only if $\tilde{w}_i(k) = 0$. Referring to the LaSalle's invariance principle [5], it can be obtained that $\tilde{w}_i(k) \to 0$. Hence, it denotes the designed observer (9.20) can well estimate the unknown model bias with its convergence proof demonstrated.

SOC tracking-based charging control with model bias compensation: In order to drive the i-th cell's SOC $x_i(k)$ $(1 \leq i \leq n)$ to track the pre-generated optimal trajectory while satisfying the safety-related constraint (9.6)–(9.8), the following SOC tracking-based charging control strategy with online compensation of model deviation is developed as follows:

$$\begin{aligned} \min_{u_i(k)} \ & \tfrac{\gamma_3}{2}(x_i(k) - x_0^r(k))^2 + \tfrac{\gamma_4}{2}u_i^2(k) \\ \text{s.t.} \ & x_i(k+1) = x_i(k) + b_i u_i(k) \\ & f_0(x_i(k+1)) + h_0(x_i(k+1))u_i(k) + \hat{w}_i(k+1) \leq y_M \\ & 0 \leq u_i(k) \leq u_M, \ \ x_i(k+1) \leq x_M \end{aligned} \tag{9.22}$$

where γ_3 and γ_4 are positive weight coefficients. The charging current $u_i(k)$ can be designed by solving the constrained optimization issue (9.22) through utilizing barrier method at each step to make the cell track its pre-generated optimal trajectory.

The proposed distributed charging control method has at least two advantages:

- First, for a battery pack consisting n serially connected cells, only n parallel optimization variables $u_i(k)$ $(1 \leq i \leq n)$ need to be computed at each step, which makes it effective for the real-time charging implementation of the battery pack. The computational burden of the battery pack charging control algorithm is reasonable, which makes it effective for practical implementation.
- Second, the charging control algorithm derived from (9.22) is closed-loop based. Since the model deviation of the cell is compensated by the observer (9.20) online, the charging control method is able to self-adjust the charging current, thus ensuring the safety-related charging constraints during charging.

9.2.4 Different Sampling Period Setting for Two Control Layers

For our designed charging control algorithm, the computational complexity of the optimal charging trajectory generation algorithm is lower with selecting a larger sampling period T, which is mainly due to the fact that the sampling step number N increases with the specified charging duration T_s for a typical fixed T since $N = \frac{T_s}{T}$. But a large T can lead to an enlarged discretization bias of the battery model for the SOC tracking-based charging control algorithm, which can adversely limit the charging performance. Hence, in order to balance the computational cost and charging performance, different sampling periods are setted for the two layers of our charging control method. To be specific, a large sampling period $T = T_1$ is adopted in the optimal average charging trajectory scheduling algorithm (9.14), while a small sampling period $T = T_2$ is utilized in the SOC tracking-based charging control method (9.22). In order to make the sampling period of the intended SOC trajectory consistent with the SOC tracking-based charging algorithm, the linear interpolation algorithm [6] is used. Following this procedure, the developed different sampling period strategy can be set to improve the performance of the proposed two-layer battery pack charging control method.

9.3 Simulation Results and Discussions

To verify the performance of the proposed user-involved charging control method with leader-followers structure, a battery pack with 10 lithium-ion cells connected in series is selected for testing. Due to the unbalanced nature of the cells in the battery pack, the cells' capacities and initial SOCs are randomly selected as shown in Table 9.1. The corresponding battery pack's initial SOC, the cells' average initial SOC and initial SOC difference are 3%, 7.5%, and 9.08%, respectively. Fig. 9.3 illustrates the mappings from the cells' SOCs to its OCVs and internal resistances, respectively. The following simplified approximate equations can be used as the parameters in the cells' nominal model in the charging control strategy that

$$
\begin{aligned}
f_0(\cdot) &= \begin{cases} c_1 SOC + d_1, & \text{if } 0 \le SOC < 5\% \\ c_2 SOC + d_2, & \text{if } 5\% \le SOC < 20\% \\ c_3 SOC + d_3, & \text{if } 20\% \le SOC < 100\% \end{cases} \\
h_0(\cdot) &= \begin{cases} c_4 SOC + d_4, & \text{if } 0 \le SOC < 5\% \\ d_5, & \text{if } 5\% \le SOC < 95\% \\ c_6 SOC + d_6, & \text{if } 95\% \le SOC < 100\% \end{cases}
\end{aligned}
\tag{9.23}
$$

with $c_1 = 3.61$, $d_1 = 3.13$, $c_2 = 1.21$, $d_2 = 3.2$, $c_3 = 0.8$, $d_3 = 3.282$, $c_4 = -0.46$, $d_4 = 0.057$, $d_5 = 0.034$, $c_6 = 2.06$, and $d_6 = -1.923$. The upper bounds of the cells' charging current and terminal voltage are selected as 3 C-rate and 4.2 V. The weight coefficients in the cost functions (9.12) and (9.22) are set to $\gamma_1 = 10^4$, $\gamma_2 = 0.1$, $\gamma_3 = 10^4$, and $\gamma_4 = 10^{-3}$. The sampling periods in the charging algorithm are chosen as $T_1 = 300$ s and $T_2 = 1$ s, respectively.

9.3.1 *Charging Results*

In the simulation, the user demanded SOC is set to $x_s = 100\%$, and the charging time is chosen as $T_s = 120$ min, respectively. Figs. 9.4a–c shows the results of the cells' SOC, charging currents, and terminal voltages under the proposed charging control

Table 9.1 Cell's capacities and initial SOCs in the battery pack

	Cell 1	Cell 2	Cell 3	Cell 4	Cell 5
Capacity	10.04Ah	9.95Ah	9.96Ah	10.05Ah	9.97Ah
Initial SOC	8%	6%	9%	12%	7%
	Cell 6	Cell 7	Cell 8	Cell 9	Cell 10
Capacity	10.01Ah	9.99Ah	10Ah	9.98Ah	10.03Ah
Initial SOC	10%	11%	3%	5%	4%

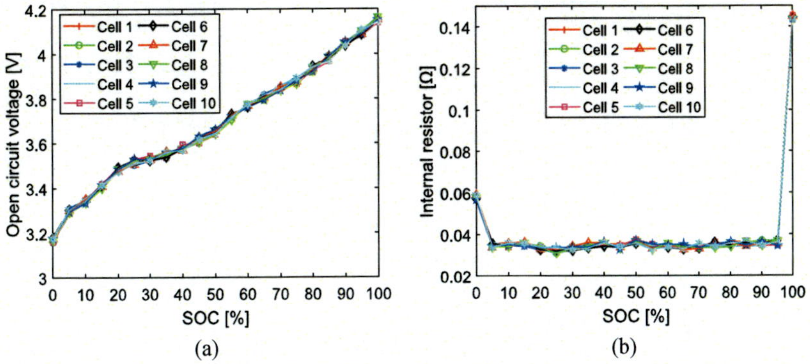

Fig. 9.3 **a** Cells' OCVs, **b** internal resistances in the battery pack

method for the battery pack. It can be seen that even if we use a simple nominal battery model rather than an accurate and complex model, the safety-related hard constraints can be still ensured within the charging process. The corresponding battery pack's SOC, cell's SOC difference, and energy loss are illustrated in Fig. 9.4d–f. The actual SOC of the battery pack is charged to 98.48%, which shows that it can well track the predetermined average charging trajectory and reach close to the user's expected SOC after charging. In addition, after charging, the cells' SOC difference in the battery pack converges from 9.08% to 0.64%, which shows the effectiveness of the multi-module charger and the proposed charging control method in cell equalization. It should be pointed out that the SOC difference among cells in the battery pack increases when these cells reach closed to the fully charged state, which mainly caused by the cells' different charging currents while there are inconsistent model parameters among the cells in the battery pack.

In order to further validate the advantages of the our designed leader-followers-based charging strategy with model bias compensation, taking the 5-th cell as an example, the detailed terminal voltages and charging currents with and without model bias compensation are measured and compared, as shown in Fig. 9.5. The 5-th cell's terminal voltage may exceed the upper limit of 4.2 V when charging without model bias compensation, while a similar constant voltage phase would be generated with on-line battery model bias compensation when the 5-th cell is nearly to be fully

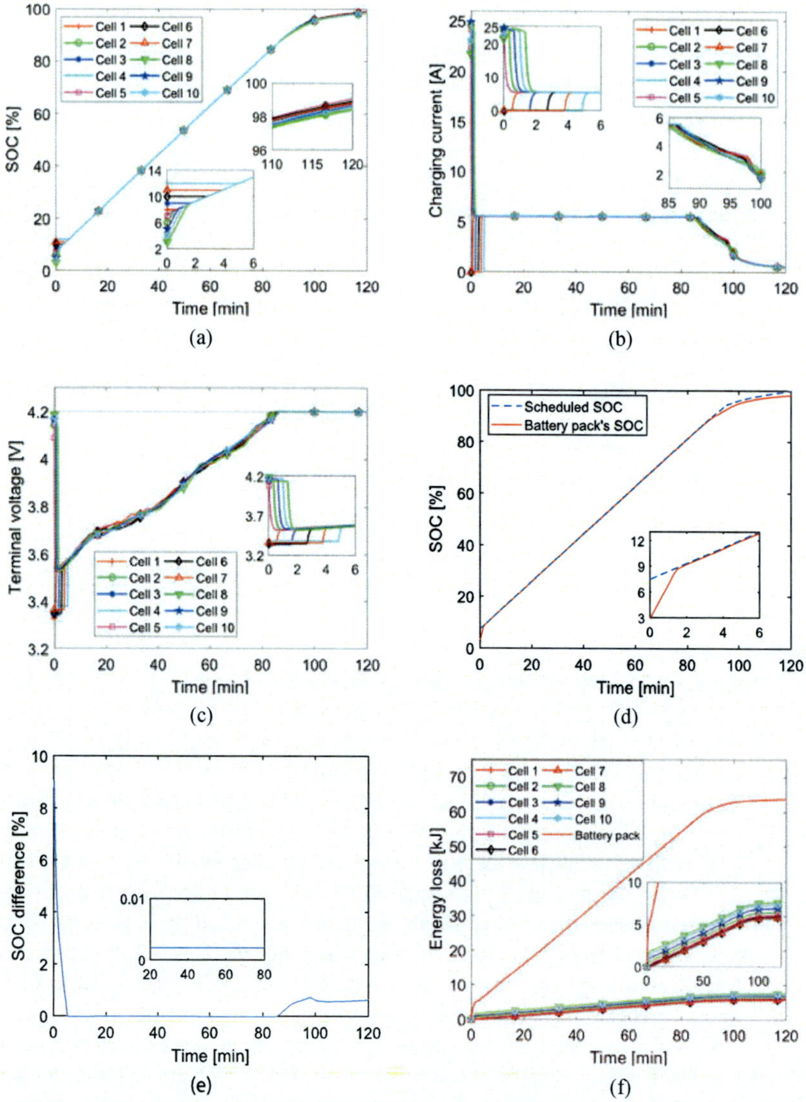

Fig. 9.4 **a** Cell's SOCs, **b** cell's terminal voltages, **c** cells' charging currents, **d** battery pack's SOC, **e** cells' SOC difference, **f** energy loss with $x_s = 100\%$ and $T_s = 120$ min

charged, suppressing the voltage rise to protect the cell's safety. As it is very difficult to accurately identify the model parameters of the battery pack in practice, our battery pack charging control method is more suitable for practical applications than other charging control methods that rely on accurate model.

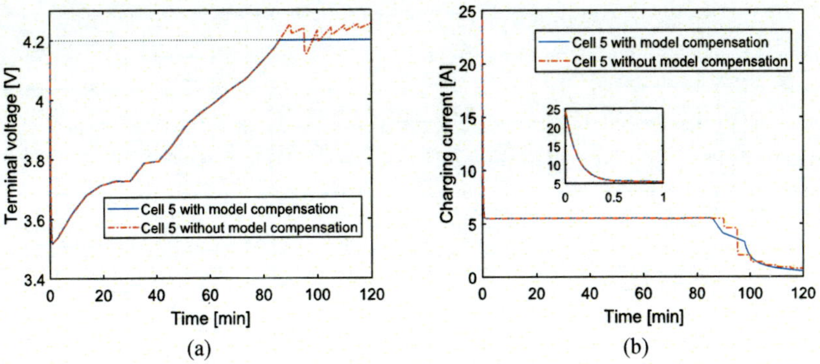

Fig. 9.5 Comparison results of the 5-th cell's **a** terminal voltages, **b** charging currents under charging control with and without model bias compensation

9.3.2 Discussions

Results for different user settings: Two types of tests are carried out for different charging durations and expected SOCs to verify the performance of the proposed charging control scheme under different user settings. First of all, for the charging duration test, the selected target SOC is $x_s = 100\%$, while the charging duration is set to $T_s = 120$ min, 90 min, 60 min, and 30 min, respectively. Then, for the expected SOC test, the charging duration is specified as $T_s = 120$ min, while the required SOC is set to $x_s = 100\%, 90\%$, and 80%, respectively. The corresponding SOC responses of the battery pack are shown in Fig. 9.6. Table 9.2 shows the results of the SOC, energy loss, and cells' SOC difference of the battery pack after charging. When the charging duration is set to 30 min, the SOC of the battery pack can be quickly charged to 77.76% with an energy loss of 134.8 kJ, and the battery pack cannot be charged to 100% within 30 min due to the charging current limitations (9.6). When the charging time is 120 min, the battery pack can also be slowly charged to 79.93% , and the corresponding energy loss is 36.5 kJ. It reflects another advantage of our designed battery pack charging control strategy, since the designed charging current can be adjusted according to user demand, which can decrease the unnecessary energy loss caused by the blind pursuit of fast charging.

Results for different weight selections: The weight coefficients γ_1 and γ_2 in the cost function (9.12) indicate the relative importance of user needs and energy loss goals. Here, we fixed $\gamma_1 = 10^4$, and then selected different γ_2 from 0.01, 0.1, 10, 100, to 1000, respectively, to verify the effect of weight selection on charging performance. Fig. 9.7 shows the results in terms of the SOC and energy loss of the battery pack. It is obvious that a larger γ_2 can result in less energy loss, but adversely affects the performance of charging the battery pack's SOC to the user's desired value. From

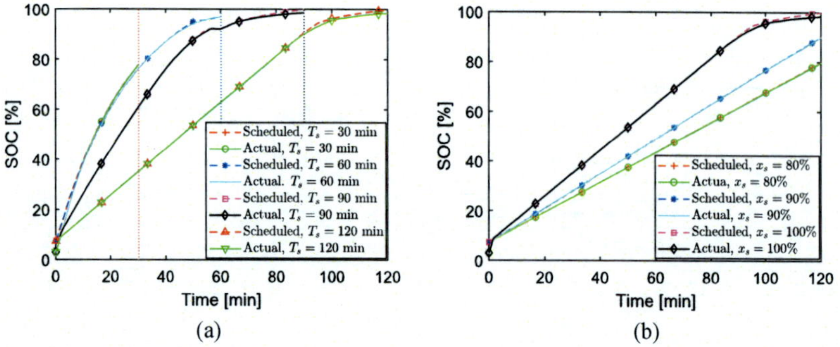

Fig. 9.6 Battery pack's SOC responses for different **a** charging durations, **b** desired SOCs

Table 9.2 Simulation results of battery pack charging for different user settings

	Battery pack's SOC (%)	Energy loss (kJ)	SOC difference (%)
$x_s = 100\%$, $T_s = 30$ min	77.76	134.8	0.03
$x_s = 100\%$, $T_s = 60$ min	96.84	142.8	0.17
$x_s = 100\%$, $T_s = 90$ min	98.62	105.6	0.61
$x_s = 100\%$, $T_s = 120$ min	98.43	63.8	0.64
$x_s = 90\%$, $T_s = 120$ min	89.93	46.1	0.002
$x_s = 80\%$, $T_s = 120$ min	79.93	36.5	0.002

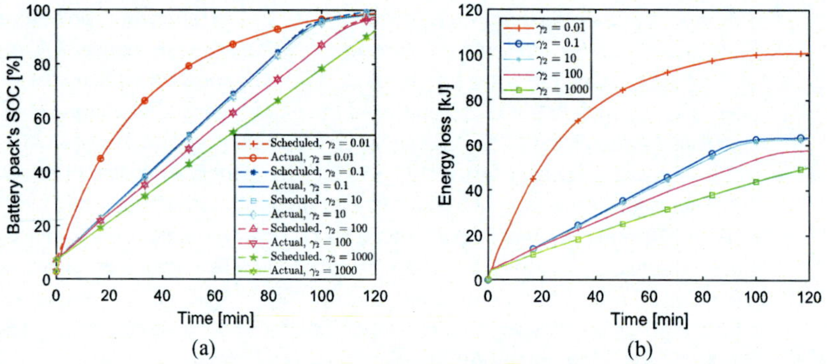

Fig. 9.7 Battery pack's **a** SOCs, **b** energy loss under the proposed charging control strategy for different selection of weight coefficients

Fig. 9.7, it is observed that $\gamma_2 = 0.1$ shows a suitable trade-off between these two conflicting objectives, and thus this weight value is selected in our proposed optimal battery pack charging control method in our study.

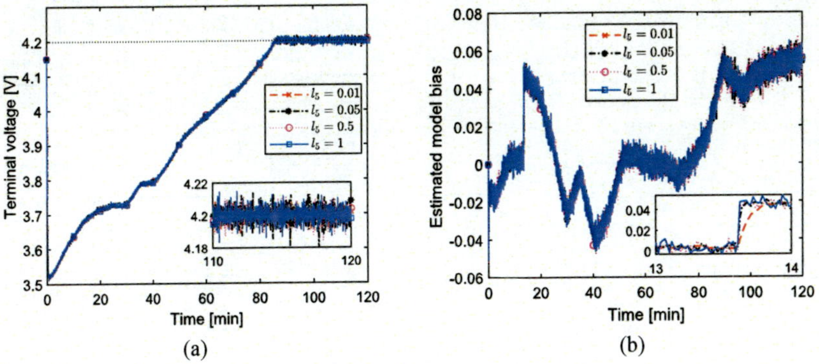

Fig. 9.8 Comparison results of the 5-th cell's responses of **a** terminal voltage, **b** estimated model bias with the selection of different observer gains

Verification with measurement noise: A Gaussian white noise with a covariance of 10^{-5} V^2 is added to the measured terminal voltage signal to verify the performance of the designed charging control method in the presence of measurement noise. Theoretically, for the designed model deviation observer, the convergence speed will be faster when the gain of the observer is larger, but this will lead to a greater amplification of the measurement noise and reduce the estimation performance. In order to select the appropriate observer gain to balance the estimation accuracy and convergence rate, many simulations are constructed with different gains, where the gain is set to 0.01, 0.05, 0.5, and 1, respectively. Fig. 9.8 shows that the simulation results in terms of the terminal voltage and estimated model deviation of the 5-th cell for different selected gain in the observer, which is consistent with the above theoretical analysis. It shows that a good charging performance of the designed charging control method is achieved that is insensitive to the selection of the observer gain, as all the selected gains can ensure the terminal voltage constraint. In practical applications, it is recommended to choose a small gain for the bias observer because the charging process usually lasts for dozens of minutes, where $l_i = 0.01$ is selected in this study.

References

1. Q. Ouyang, Z. Wang, K. Liu, G. Xu, and Y. Li, "Optimal charging control for lithium-ion battery packs: A distributed average tracking approach," *IEEE Transactions on Industrial Informatics*, vol. 16, no. 5, pp. 3430-3438, 2020.
2. K. Liu, X. Hu, Z. Yang, Y. Xie, and S. Feng, "Lithium-ion battery charging management considering economic costs of electrical energy loss and battery degradation," *Energy Conversion and Management*, vol. 195, pp. 167-179, 2019.
3. C. Yan, H. Fang, and H. Chao, "Battery-aware time/range-extended leader-follower tracking for a multi-agent system," in *American Control Conference*, 2018, pp. 3887-3893.

4. S. Boyd, L. Vandenberghe, *Convex Optimization*, Cambridge University Press, New York, NY, USA, 2004.
5. H. K. Khalil, *Nonlinear Systems*, 3rd ed. Englewood Cliffs, NJ, USA: Prentice-Hall, 2002.
6. M. Hazewinkel, *Linear Interpolation*, 1st ed. Netherlands: Springer, 2001.

Chapter 10
Fast Battery Charging Control for Battery Packs

The multi-module charger utilized in Chaps. 8 and 9 can increase the cost compared with traditional battery chargers, especially for the battery pack consisting of a large amount of serially connected cells, as this charger is composed of many small charger modules for all cells. By using the combined battery pack charging system combining the traditional charger and bidirectional cell-to-cell equalizers, it can enables cell equalization without additional cost of the charger in practical applications.

Based on this battery pack charging system, in this chapter, an optimization-based fast charging control strategy is proposed with considering the multiple objectives in terms of charging time, charging energy loss, and charging constraints to coordinate and optimize the current provided by the charger and the equalizing currents [1]. Next, a two-layer optimization algorithm is formulated to solve it, where the top layer is a binary-search-based charging time region contraction method to determine the minimum expected charging time, and the bottom layer utilizes the barrier method to get the corresponding optimal charging and equalizing currents that can make the difference between the actual and desired SOCs of the battery pack within the expected charging time meet the pre-set tolerance requirement.

10.1 Charging Model for the Battery Pack

For a battery pack, the charging procedure has to be terminated when one of the cells is fully charged to avoid its overcharging, which results in the insufficient charging of the entire battery pack due to the phenomenon of cell imbalance in the battery pack [2]. To remedy this deficiency, a novel battery pack charging system is developed here by integrating the traditional charger and bidirectional cell-to-cell equalizers, as shown in Fig. 10.1. It should be pointed out that this charging system will not increase the hardware expense since the charger and cell equalizers are usually equipped in the BMS though they are not combined as in the proposed way in practice. Since the

Q. Ouyang and J. Chen, *Advanced Model-Based Charging Control for Lithium-Ion Batteries*, https://doi.org/10.1007/978-981-19-7059-7_10

literature [3] has already studied the implementation of chargers and bidirectional cell-to-cell equalizers, this work will mainly focus on the coordinated optimization control of the currents provided by the charger and equalizers.

10.1.1 Charging Current Model

As illustrated in Fig. 10.1, for a battery pack consisting of serially connected n cells, energy can be transferred between the i-th and $(i+1)$-th $(1 \leq i \leq n-1)$ cells in bidirectional directions through their connected equalizer i, where the equalizing currents are defined as $I_{el_i}(k)$ and $I_{er_i}(k)$, respectively. As the energy needs to be transferred from the cells with higher SOC to the cells with lower SOC during within the charging period to achieve cell equalization of the battery pack, the equalizing current direction through the i-th equalizer can be predetermined by comparing the i-th and $(i+1)$-th cells' SOCs to reduce the number of control variables. In light of this, $I_{el_i}(k)/I_{er_i}(k)$ is selected as the controlled equalizing current and replaced by a new notation $I_{e_i}(k)$, when the i-th/$(i+1)$-th cell's initial SOC is larger than that of the $(i+1)$-th/i-th cell. Hence, the cells' equalizing currents can be represented as follows [4]:

$$\begin{aligned} I_{el_i}(k) &= (-k_i + k_i'\beta_i)I_{e_i}(k) \\ I_{er_i}(k) &= (k_i\beta_i - k_i')I_{e_i}(k) \end{aligned} \tag{10.1}$$

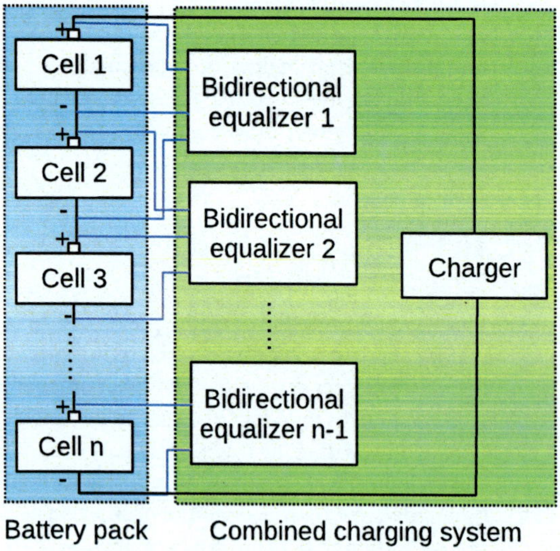

Fig. 10.1 Combined charging system for the battery pack

with

$$\begin{cases} k_i = 1, k_i' = 0, \text{ if } SOC_i(0) \geq SOC_{i+1}(0) \\ k_i = 0, k_i' = 1, \text{ if } SOC_i(0) < SOC_{i+1}(0) \end{cases}$$

where $I_{e_i}(k)$ is the controlled equalizing current of the i-th equalizer, $SOC_i(0)$ and $SOC_{i+1}(0)$ denotes the initial SOCs of the i-th and $(i+1)$-th cells, respectively; β_i $(0 < \beta_i < 1)$ is the ratio between the equalizing currents $I_{el_i}(k)$ and $I_{er_i}(k)$, which can be defined as the energy transfer efficiency of the i-th equalizer. Note that $I_{e_i}(k)$ is set as a positive variable, and $I_{el_i}(k)$ and $I_{er_i}(k)$ are positive for charging and negative for discharging.

The i-th cell's total charging current includes the current provided by the battery charger and the equalizing currents from its adjacent cells. Based on (10.1), the i-th cell's total charging current $I_{B_i}(k)$ can be calculated as:

$$\begin{aligned} I_{B_1}(k) &= I_c(k) + (-k_1 + k_1'\beta_1)I_{e_1}(k) \\ I_{B_i}(k) &= I_c(k) + (k_{i-1}\beta_{i-1} - k_{i-1}')I_{e_{i-1}}(k) \\ &\quad + (-k_i + k_i'\beta_i)I_{e_i}(k) \qquad (2 \leq i \leq n-1) \\ I_{B_n}(k) &= I_c(k) + (k_{n-1}\beta_{n-1} - k_{n-1}')I_{e_{n-1}}(k) \end{aligned} \tag{10.2}$$

where $I_c(k)$ denotes the i-th cell's current provided by the battery charger.

10.1.2 Battery Pack Model

Here, the Rint model as shown in Fig. 3.5 is adopted to characterize the dynamics of each cell in the battery pack, as it can strike a balance between computational complexity and model accuracy. The Rint model is composed of a voltage source that simulates the energy storage and a serially connected internal resistance that characterizes the energy loss during charging. The electrical governing equations of the i-th $(1 \leq i \leq n)$ cell in the battery pack can be derived as

$$\begin{aligned} SOC_i(k+1) &= SOC_i(k) + \tfrac{\eta_0 T}{3600 Q_i} I_{B_i}(k) \\ V_{B_i}(k) &= V_{OC_i}(k) + R_{0_i}(k) I_{B_i}(k) \end{aligned} \tag{10.3}$$

where $SOC_i(k)$, Q_i, and $V_{B_i}(k)$ represents the i-th cell's SOC, capacity, and terminal voltage of the i-th cell, respectively; η_0 and T are the Coulombic efficiency and sampling period; $V_{OC_i}(k) = f_i(SOC_i(k))$ and $R_{0_i}(k) = h_i(SOC_i(k))$ denote the SOC-dependent OCV and internal resistance of the i-th cell's, respectively.

Due to the existence of energy imbalance among cells, the cell with the lowest SOC limits the available capacity of the whole battery pack. Therefore, the battery pack's SOC can be defined as the lowest SOC of the cells in the battery pack, which can be expressed as

$$SOC_p(k) = min\{SOC_1(k), \cdots, SOC_n(k)\} \tag{10.4}$$

where $SOC_p(k)$ represents the battery pack's SOC. From (10.2)-(10.4), the charging model of the battery pack can be written as:

$$\begin{aligned} x(k+1) &= x(k) + D(1_n u_1(k) + B_1 u_2(k)) \\ y(k) &= f(x(k)) + h(x(k))(1_n u_1(k) + B_1 u_2(k)) \\ z(k) &= \min\{x_1(k), \cdots, x_n(k)\} \end{aligned} \tag{10.5}$$

with

$$\begin{aligned} x(k) &= [x_1(k), \cdots, x_n(k)]^T \triangleq [SOC_1(k), \cdots, SOC_n(k)]^T \\ y(k) &= [y_1(k), \cdots, y_n(k)]^T \triangleq [V_{B_1}(k), \cdots, V_{B_n}(k)]^T \\ z(k) &\triangleq SOC_p(k) \end{aligned}$$

where the control variables are $u_1(k) \triangleq I_c(k) \in \mathbb{R}$ and $u_2(k) = [u_{2_1}(k), \cdots, u_{2_n}(k)]^T \triangleq [I_{e_1}(k), \cdots, I_{e_n}(k)]^T \in \mathbb{R}^{n-1}$, respectively; 1_n represents an $n \times 1$ column vector of ones, $f(\cdot) = [f_1(\cdot), \cdots, f_n(\cdot)]^T \in \mathbb{R}^n$; $D = diag\{\frac{\eta_0 T}{3600 Q_1}, \cdots, \frac{\eta_0 T}{3600 Q_n}\} \in \mathbb{R}^{n \times n}$, and $h(\cdot) = diag\{h_1(\cdot), \cdots, h_n(\cdot)\} \in \mathbb{R}^{n \times n}$ with $diag\{\cdot\}$ the diagonal matrix; $B_1 \in \mathbb{R}^{n \times n-1}$ is

$$B_1 = \begin{bmatrix} -k_1 + k_1'\beta_1 & 0 & \cdots & 0 \\ k_1\beta_1 - k_1' & -k_2 + k_2'\beta_2 & \cdots & 0 \\ \cdots & \cdots & \cdots & \cdots \\ 0 & 0 & \cdots & k_{n-1}\beta_{n-1} - k_{n-1}' \end{bmatrix}.$$

10.2 Control Objectives and Constraints

The charging control objectives for the battery pack are firstly formulated, and then, the constraints of the battery pack and the combined charging system are listed.

10.2.1 Charging Objectives

Charging time: Fast charging speed is one of the most important goals for the battery pack charging management, since long charging time leads to inconvenience use of the battery pack and eventually results in the consumers' anxiety. The control strategy aims to minimize the consumed time for the battery pack to be charged from an initial SOC vector of $x(0)$ to the desired value of $x_d 1_n$. The corresponding cost function in terms of the charging time can be straightforwardly constructed as

$$J_t = NT \tag{10.6}$$

where J_t represents the charging time and N is the sampling step number with $x(N) = x_d 1_n$.

Charging energy loss: Another important objective commonly considered is the energy loss during the charging procedure. In practice, less energy loss of the battery pack is desired to improve the charging efficiency. From Eq. (10.5), the corresponding cost function with respect to charging energy loss can be formulated as

$$\begin{aligned} J_e = \sum_{k=0}^{N-1} & T(1_n u_1(k) + B_1 u_2(k))^T h(x(k+1)) \\ & \times (1_n u_1(k) + B_1 u_2(k)) \end{aligned} \tag{10.7}$$

where J_e denotes the energy loss of the battery pack during charging. Note that by constraining the cost function (10.7), the charging current can be suppressed, thus restraining the temperature rise of the battery pack.

10.2.2 Charging Constraints

During the charging process, The phenomena, such as overcharging, overcurrent, and overvoltage, can cause the capacity degradation acceleration of the battery pack and even lead to safety problems. From this view, hard constraints including the cells' SOCs, charging currents, and terminal voltages in the battery pack has to be taken into consideration as:

$$\begin{aligned} & x(k+1) \le x_M 1_n \\ & 1_n u_1(k) + B_1 u_2(k) \le u_M 1_n \\ & f(x(k+1)) + h(x(k+1))(1_n u_1(k) + B_1 u_2(k)) \le y_M 1_n \end{aligned} \tag{10.8}$$

where $x_M \in \mathbb{R}$, $u_M \in \mathbb{R}$, and $y_M \in \mathbb{R}$ are the upper bounds of the cells' SOCs, charging current and terminal voltage, respectively. Considering some physical circuit limitations of the charger and equalizers, the following constraints are then imposed:

$$\begin{aligned} & 0 \le u_1(k) \le u_{1_M} \\ & 0_{n-1} \le u_2(k) \le u_{2_M} 1_{n-1} \end{aligned} \tag{10.9}$$

where $u_{1_M} \in \mathbb{R}$ and $u_{2_M} \in \mathbb{R}$ stand for, respectively, the maximum charging and equalizing currents that the charger and equalizers can provide; 0_{n-1} is the column vector with $n-1$ zeros.

10.3 Fast Charging Control Strategy Design

10.3.1 Charging Control Algorithm Formulation

A high-quality fast battery pack charging control strategy should pursue both short charging time and low charging energy loss. But these two objectives are in conflict since shorter charging time causes higher energy loss. To handle these two conflicting objectives, since the charging time target is of relatively higher importance than the energy loss objective in practical applications, a bounded objective function method [5] is adopted here to convert them into a single objective by redefining the charging energy loss objective into a constraint as

$$J_e \leq J_{eM} \tag{10.10}$$

with J_{eM} the selected upper bound of the battery pack's charging energy loss. Based on the above analysis, the fast charging control strategy for the battery pack can be summarized as the following constrained optimization problem:

$$\begin{aligned}
&\min_{u_1(0),u_2(0),\cdots,u_1(N-1),u_2(N-1)} \quad J_t \\
&\text{s.t. } x(k+1) = x(k) + D(1_n u_1(k) + B_1 u_2(k)), \quad x(0) \\
&\quad f(x(k+1)) + h(x(k+1))(1_n u_1(k) + B_1 u_2(k)) \leq y_M 1_n \\
&\quad 1_n u_1(k) + B_1 u_2(k) \leq u_M 1_n, \quad x(k+1) \leq x_M 1_n \\
&\quad 0 \leq u_1(k) \leq u_{1_M}, \quad 0_{n-1} \leq u_2(k) \leq u_{2_M} 1_{n-1} \\
&\quad J_e \leq J_{eM}, \quad x(N) = x_d 1_n.
\end{aligned} \tag{10.11}$$

Note that (10.11) is difficult to be directly solved, as there exists terminal state constraint $x(N) = x_d 1_n$ and the terminal time NT is not fixed. In addition, the relationship between the charging time and charging current cannot be explicitly expressed, which makes traditional numerical optimization methods, such as the gradient-based methods, ineffective to solve the considered problem.

To effectively obtain the optimal solution of (10.11), a two-layer optimization algorithm is designed here, which continues to narrow the charging time range until the shortest expected charging time is determined in the top layer and utilizes a gradient-based method to calculate the corresponding optimal charging and equalizing currents in the bottom layer.

10.3.2 Two-Layer Optimization Algorithm

The designed two-layer optimization method is described in detail as follows:

- In the bottom layer, for an expected charging time J_t given by the top layer, the terminal state constraint in (10.11) can be transformed to design the optimal charging

and equalizing currents $u_1(k)$ and $u_2(k)$ ($0 \le k \le N-1$) to minimize the difference between the terminal state $x(N)$ and the desired state $x_d 1_n$. Through such operation, (10.11) can be transformed to a conventional constrained optimization issue that can be effectively solved by the gradient-based algorithms such as the barrier method [6].
- In the top layer, a binary search algorithm is developed, where the charging time region is contracted from an initial set range until the minimum expected charging time is obtained that can make the difference between the actual and desired states meet the pre-set accuracy in the bottom layer. Therefore, the optimal charging and equalizing currents can be obtained that leads to the shortest charging period while guaranteeing the charging constraints in (10.11).

Bottom layer: charging and equalizing currents optimization: In the bottom layer, the sampling step number N is fixed to ensure the computational burden of the bottom layer control algorithm consistent for different charging times. In other words, for an expected charging time J_t given by the top layer, the sampling period is calculated as $T = \frac{J_t}{N}$. To simplify the notations, we define $u(k) = [u_1(k), u_2^T(k)]^T \in \mathbb{R}^n$. From (10.5), the state vector of the battery pack can be updated as

$$x(k+1) = x(k) + DBu(k) \tag{10.12}$$

where $B = [1_n, B_1] \in \mathbb{R}^{n \times n}$. As described above, the optimization issue with terminal state constraint $x(N) = x_d 1_n$ in (10.11) is transformed to drive the battery pack's terminal state $x(N)$ toward to the desired one $x_d 1_n$ to the largest extent as

$$\begin{aligned} \min_{u(0),u(1),\cdots,u(N-1)} \quad & (x(N) - x_d 1_n)^T (x(N) - x_d 1_n) \\ \text{s.t.} \quad & x(k+1) = x(k) + DBu(k), \ \ x(0) \\ & f(x(k+1)) + h(x(k+1))Bu(k) \le y_M 1_n \\ & Cu(k) \le u_L, \ \ x(k+1) \le x_M 1_n \\ & J_e \le J_{eM}, \ \ NT = J_t \end{aligned} \tag{10.13}$$

where $C = [B^T, I_n, -I_n]^T \in \mathbb{R}^{3n \times n}$, $u_L = [u_M 1_n^T, u_{1_M}, u_{2_M} 1_{n-1}^T, 0_n^T]^T \in \mathbb{R}^{3n}$, and I_n denotes an identity matrix with dimensions of $n \times n$. With defining $U = [u^T(0), \cdots, u^T(N-1)]^T \in \mathbb{R}^{nN}$, (10.12) can be rewritten as follows:

$$x(k) = x(0) + DH_k U \tag{10.14}$$

where $H_k \triangleq [\Upsilon_k, \Theta_k] \in \mathbb{R}^{n \times nN}$ with $\Upsilon_k = [B, \cdots, B] \in \mathbb{R}^{n \times kn}$ and $\Theta_k = [0_{n \times n}, \cdots, 0_{n \times n}] \in \mathbb{R}^{n \times (N-k)n}$. Then, (10.13) can be rewritten as the following standard constrained optimization issue:

$$\begin{aligned} \min_{U} \quad & J_1(U) \\ \text{s.t.} \quad & F(U) + G(U)U \leq Y_M \\ & MU \leq X_C, \; \Phi U \leq U_M \\ & TU^T\Gamma(U)U \leq J_{eM} \end{aligned} \tag{10.15}$$

where $J_1(U) \in \mathbb{R}$, $F(U) \in \mathbb{R}^{nN}$, $G(U) \in \mathbb{R}^{nN\times nN}$, $X_C \in \mathbb{R}^{nN}$, $Y_M \in \mathbb{R}^{nN}$, $\Phi \in \mathbb{R}^{3nN\times nN}$, $M \in \mathbb{R}^{nN\times nN}$, $U_M \in \mathbb{R}^{3nN}$, and $\Gamma(U) \in \mathbb{R}^{nN\times nN}$, respectively, are

$$\begin{aligned} J_1(U) &= U^T H_N^T D^T D H_N U + 2(x(0) - x_d 1_n)^T D H_N U \\ &\quad + (x(0) - x_d 1_n)^T (x(0) - x_d 1_n) \\ F(U) &= [f^T(x(0) + DH_1U), \cdots, f^T(x(0) + DH_NU)]^T \\ G(U) &= diag\{h(x(0) + DH_1U)B, \cdots, h(x(0) + DH_NU)B\} \\ X_C &= [(x_M 1_n - x(0))^T, \cdots, (x_M 1_n - x(0))^T]^T \\ Y_M &= y_M 1_{nN}, \quad \Phi = diag\{C, \cdots, C\} \\ M &= [(DH_1)^T, \cdots (DH_N)^T]^T, \quad U_M = [u_L^T, \cdots, u_L^T]^T \\ \Gamma(U) &= diag\{B^T h(x(0) + DH_1U)B, \cdots, B^T h(x(0) + DH_NU)B\}. \end{aligned}$$

Note that the problem in (10.15) can be readily solved using some traditional gradient-based optimization methods, e.g., the barrier method in [6]. The solution will offer the optimal charging current sequence over the overall charging time horizon.

Top layer: charging time optimization: The terminal state constraint $x(N) = x_d 1_n$ in (10.11) has been converted to minimize the difference between $x(N)$ and $x_d 1_n$ as in (10.15) for an expected charging time J_t in the bottom layer. But the cells' actual SOCs when charging is complete may deviate significantly from the desired one when the expected charging time J_t given to the bottom layer is too short. It will make the charging result fail to meet the user's demand. On the other hand, if the control input vector U designed by the bottom layer can make the root mean square (RMS) of the difference between the terminal and desired SOCs satisfy

$$\tfrac{1}{\sqrt{n}}||x(N) - x_d 1_n|| \leq \epsilon_1 \tag{10.16}$$

with ϵ_1 being a pre-set tolerance and $|| \cdot ||$ denoting the 2-norm, it indicates that the choice of J_t may be too large. Otherwise, it denotes that the given expected charging time from the top layer J_t is less than the minimum required one.

Inspired by this, a binary-search-based charging time region contraction algorithm is proposed here, where the charging time region is narrowed at each iteration step from an initial region $[T_{c1_0}, T_{c2_0}]$ until the minimum expected charging time J_{t_m} is determined. For the k-th iteration step, the middle value of the region in the $(k-1)$-th iteration $\lambda_k = \frac{T_{c1_{k-1}} + T_{c2_{k-1}}}{2}$ is chosen as the expected charging time for the bottom-layer optimization, where $[T_{c1_{k-1}}, T_{c2_{k-1}}]$ is the charging time region after the $(k-1)$-th iteration. The charging time region can be updated by the rule as follows:

- If (10.16) can be satisfied by solving (10.15) with the given expected charging time $J_t = \lambda_k$ in the bottom layer, it means that the right end point of the charging time region $T_{c2_{k-1}}$ is selected a bit large and it should be replaced by the middle value λ_k, i.e., the charging time region after the k-th iteration is $[T_{c1_{k-1}}, \lambda_k]$.
- Otherwise, it indicates that the charging time $T_{c1_{k-1}}$ is not enough to complete the charging task. Therefore, the middle value λ_k should be used to replace the left end point of the charging time region $T_{c1_{k-1}}$, and the new charging time region becomes $[\lambda_k, T_{c2_{k-1}}]$.

It should be pointed out that every iteration will halve the charging time region by 50%, which can lead to a fast convergence to the target value, i.e., the expected minimum charging time. The iteration process will be terminated and the minimum charging time can be obtained as

$$J_{t_m} = T_{c2_k} \ \text{ if } \ |T_{c2_k} - T_{c1_k}| \leq \epsilon_2 \tag{10.17}$$

where ϵ_2 denotes the pre-set tolerance. Therefore, the corresponding optimal charging current can be computed by solving (10.15) with the obtained J_{t_m}. Note that smaller ϵ_1 and ϵ_2 can lead to a higher accuracy of charging time prediction but cause more computation complexity. Suitable ϵ_1 and ϵ_2 can be assigned based on the user's demand in practical applications.

The detailed description of the proposed two-layer optimization algorithm is presented as Algorithm 10.1.

Algorithm 10.1: Two-layer optimization method

---

(1) Set the cells' initial and target SOC vectors $x(0)$ and $x_d 1_n$, and select the initial charging period region $[T_{c1_0}, T_{c2_0}]$.

(2) For the k-th iteration, set the intermediate variable $\lambda_k = \frac{T_{c1_{k-1}} + T_{c2_{k-1}}}{2}$ and choose it as the expected charging time for (10.15) as $J_t = \lambda_k$.

(3) Use the barrier method [6] to solve (10.15). If there exists a solution of (10.15) that can make (10.16) satisfied, select $T_{c1_k} = T_{c1_{k-1}}$ and $T_{c2_k} = \lambda_k$. Otherwise, set $T_{c1_k} = \lambda_k$ and $T_{c2_k} = T_{c2_{k-1}}$.

(4) Stop and output the optimal solution of (10.15) with the expected charging time $J_{tm} = T_{c2_k}$, if $|T_{c2_k} - T_{c1_k}| \leq \epsilon_2$. Otherwise, set $k + 1 \rightarrow k$ and return to Step 2).

10.4 Simulation Results

To validate the effectiveness of the designed fast battery pack charging control strategy, MATLAB-based simulations are carried out for a battery pack consisting of 10 serially connected cells with the nominal capacity and nominal voltage

Table 10.1 Cells' initial SOCs in the battery pack

$x_1(0)$	$x_2(0)$	$x_3(0)$	$x_4(0)$	$x_5(0)$	$x_6(0)$	$x_7(0)$	$x_8(0)$	$x_9(0)$	$x_{10}(0)$
15%	1%	20%	27%	8%	5%	26%	11%	12%	14%

of 2.1 Ah and 3.7V. The cells' OCVs and internal resistances are the same as selected in [7]. The upper limitations of the cells' SOC, charging current, and terminal voltage are set to 100%, 3 C-rate, and 4.2 V, respectively. The maximum charging current that the charger can provide is 6 A, and the maximum equalizing current of the equalizers is 1 A, respectively. The energy transfer efficiency of the equalizers is set to 0.9. The maximum allowed energy loss of the battery pack is selected as 5% of its total recharged energy and it can be calculated as $J_{eM} = 5\% \times 3.7 \times (x_d 1_n - x(0))^T \times 2.1 \times 1_n \times 3600$ J. The initial charging time region is set to [10 min, 360 min]. The tolerances ϵ_1 and ϵ_2, and the sampling step number are set as 0.5%, 1 min, and $N = 10$, respectively. The desired SOC of the battery pack is set to $x_d = 100\%$. The initial SOCs of the cells are selected as in Table 10.1, where the battery pack's initial SOC is 1%.

The results in terms of the battery pack's SOC and energy loss under the proposed charging control strategy are shown in Fig. 10.2a, which denotes that the battery pack's SOC can be charged from 1% to 99.12% within 40.76 min with the energy loss of 12,042 J exactly satisfying the pre-set limitation J_{eM}. Figure 10.2b and c illustrates the control variables, i.e., the charging current provided by the charger and the controlled equalizing currents. The cells' SOCs in the battery pack are shown in Fig. 10.2d, where it can be observed that they all reach the range of [99.12%, 100%] after charging. The results of the total charging currents and terminal voltages of the cells are illustrated in Fig. 10.2e–f, which shows that their constraints in (10.8) can be well satisfied. From the above simulation results, it can be concluded the designed optimal fast charging method can drive the SOC of the battery pack to its desired value within a shortest charging period while ensuring the safety-related charging constraints.

Comparison with traditional chargers: For the traditional chargers, all cells in the battery pack are charged with the same current due to their series connection, which lacks the ability of mitigating the cell imbalance. If they are applied on the adopted battery pack, the charging procedure must be terminated when the 4-th cell is fully charged ($SOC_4 = 100\%$) to avoid its overcharging. However, the battery pack's SOC is only 74% restricted by the second cell. It means that the battery pack is insufficiently charged. This comparison shows the superior performance of our strategy in enhancing the battery pack's effective capacity.

Comparison with CC-CV method with constant equalizing currents: To further validate the superiority of the designed battery pack charging control method, the simulation results of CC-CV method with constant equalizing currents are also provided here as the comparison. The constant current and constant voltage in the CC-CV method are set to 1 C-rate and 42 V for the battery pack, respectively. The

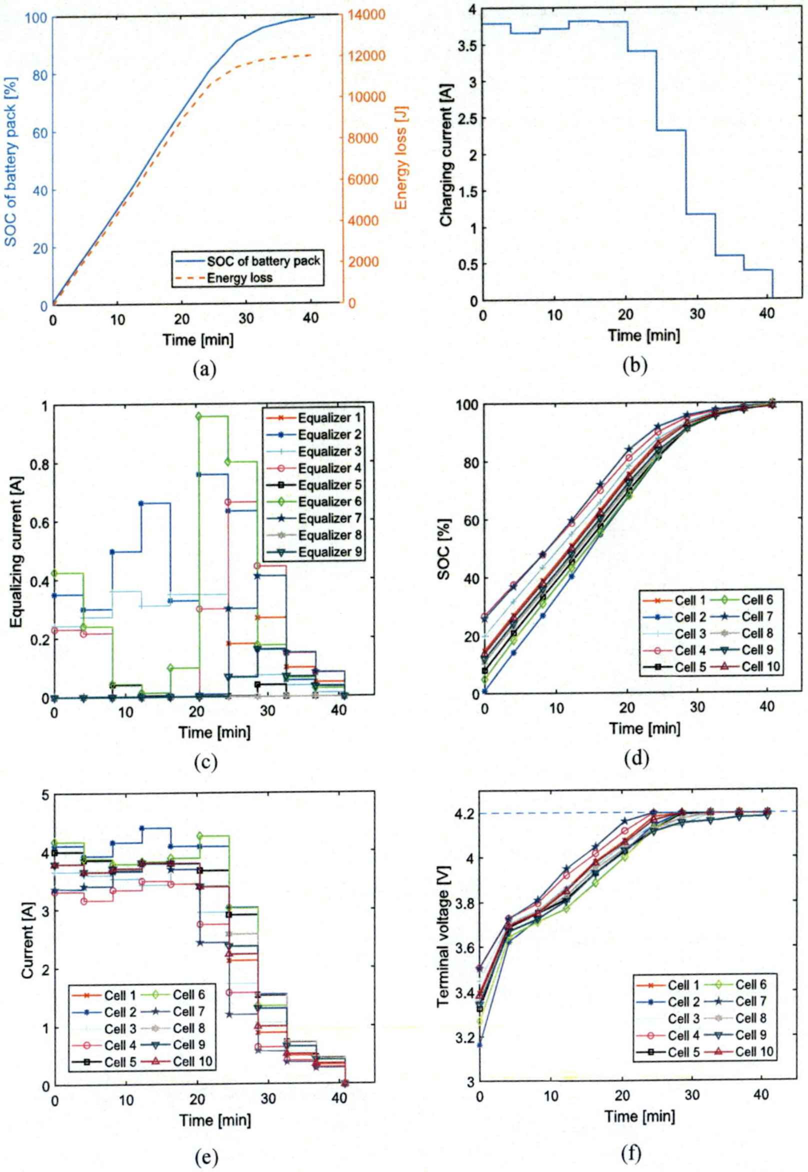

Fig. 10.2 Simulation results of **a** battery pack's SOC and energy loss responses, **b** current provided by the charger, **c** controlled equalizing currents, **d** cells' SOCs, **e** cells' currents, **f** cells' terminal voltages

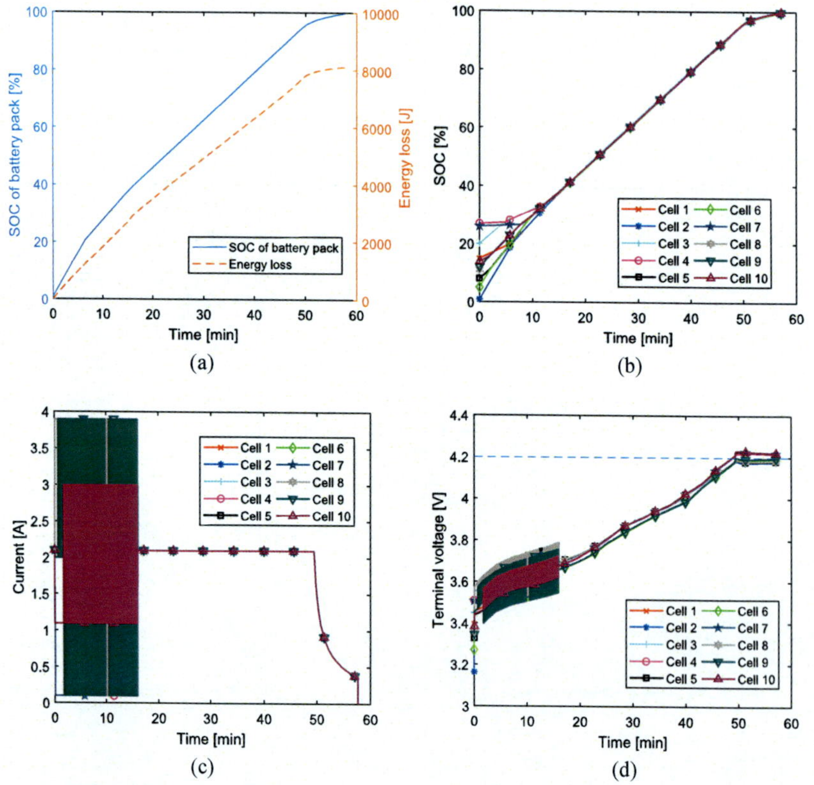

Fig. 10.3 Simulation results for CC-CV method with constant equalizing currents in terms of **a** battery pack's SOC and energy loss responses, cells' **b** SOCs, **c** currents, and **d** terminal voltages

equalizing current is set to 1 A, and the equalizing procedure is terminated when the RMS of the cells' SOC difference satisfies $\frac{1}{\sqrt{n}}||x(k) - \frac{1}{n}1_n 1_n^T x(k)|| \leq 0.1\%$. The corresponding simulation results in terms of the battery pack's SOC and energy loss are shown in Fig. 10.3a. It can be observed that the consumed charging time is 57.63 min, and the energy loss of the battery pack is 8, 122 J. Figure 10.3b–d illustrates the cell's SOCs, currents, and terminal voltages in the battery pack, respectively. It shows that the limitation of the cells' terminal voltages cannot be guaranteed without the combined optimization of the charger and equalizers, which is because of the existence of the difference in the cells' OCVs and internal resistances. In addition, the CC-CV method with constant current of 2 C-rate and 3 C-rate are also carried out here, with the simulation results illustrated in Table 10.2. Note that as the energy loss optimization is not considered in these CC-CV methods, they cannot be as effective as our designed charging control strategy that can minimize the charging time while limiting the energy loss of the battery pack below the pre-set bound.

Table 10.2 Comparison results with CC-CV method with constant equalizing currents

	Charging time (min)	Energy loss (J)
Proposed method	40.76	12,042
CC-CV with 1-C rate, 1A equalizing current	57.63	8122
CC-CV with 2-C rate, 1A equalizing current	34.17	14,491
CC-CV with 3-C rate, 1A equalizing current	28.2	20,284

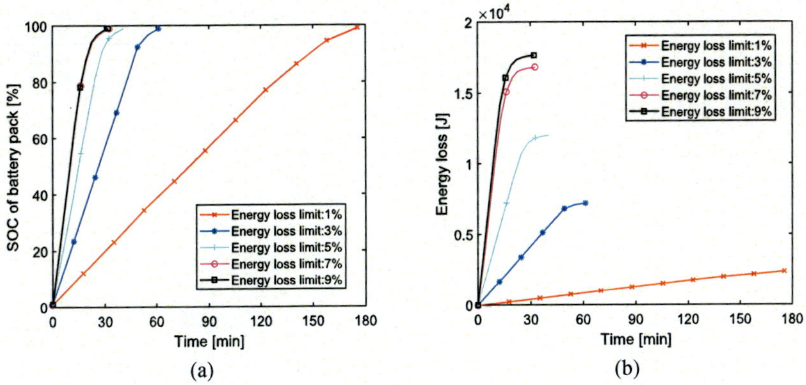

Fig. 10.4 **a** SOC response, **b** energy loss of the battery pack for different selected energy loss limitations

Different energy loss limitations: To validate the impact of the selected energy limitation on the charging performance, four more simulations are carried out with the maximum allowed energy loss set to 1, 3, 7, and 9% of its recharged energy, respectively. The results of the battery pack's SOC and energy loss are illustrated in Fig. 10.4 and Table 10.3, which shows that the objectives of charging time and energy loss are conflicting, i.e., a shorter charging time leads to more consumed energy loss. However, the designed optimal battery pack charging control strategy can minimize the charging time while constraining the energy loss of the battery pack below the pre-set limit, which thereby enables a excellent balance between these two conflicting objectives. From the comparison results, 5% of its recharged energy shows a suitable trade-off between these two objectives. Hence, it is chosen in our charging control strategy. Note that users can set the energy loss limit based on their actual demands in practice.

Different cells' initial SOCs: Another four tests are conducted on the battery pack with different cells' initial SOC vectors, showing in Cases 1–4 as follows:

- Case 1: $x(0) = [21\%, 18\%, 3\%, 12\%, 24\%, 17\%, 13\%, 7\%, 25\%, 15\%]^T$;
- Case 2: $x(0) = [17\%, 1\%, 18\%, 6\%, 8\%, 22\%, 13\%, 7\%, 21\%, 5\%]^T$;
- Case 3: $x(0) = [15\%, 7\%, 16\%, 20\%, 11\%, 27\%, 9\%, 5\%, 26\%, 10\%]^T$;
- Case 4: $x(0) = [16\%, 5\%, 19\%, 20\%, 12\%, 10\%, 25\%, 8\%, 11\%, 17\%]^T$.

Table 10.3 Comparison results for different energy loss limits

Energy loss limit	Charging time (min)	Energy loss (J)	Terminal SOC (%)
1% of its recharged energy	175.4	2408	99.07
3% of its recharged energy	61.27	7225	99.03
5% of its recharged energy	40.76	12,042	99.12
7% of its recharged energy	32.56	16,859	99.16
9% of its recharged energy	31.88	17,666	99.12

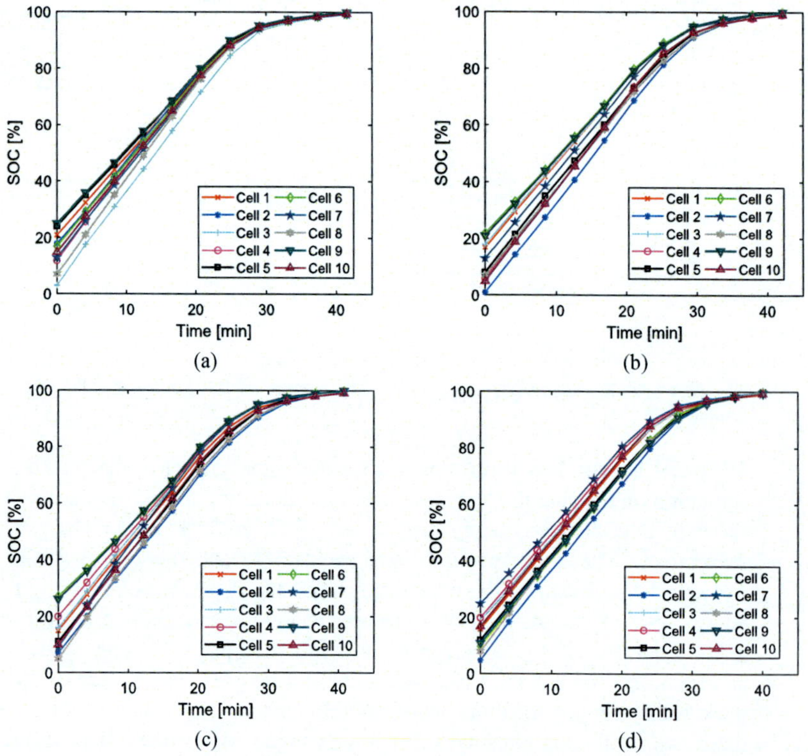

Fig. 10.5 Cells' SOC response for the battery pack with the initial SOCs as in **a** Case 1, **b** Case 2, **c** Case 3, **d** Case 4

The corresponding results in terms of the cells' SOC responses are shown in Fig. 10.5.The comparison results of the consumed charging time, energy loss, and terminal SOC of the battery pack for different cells' initial SOCs are presented in Table 10.4. They demonstrate the good performance of our designed optimal fast battery pack charging control method for all cases.

Table 10.4 Simulation results for different cells' initial SOCs

Cells' initial SOCs	Charging time (min)	Energy loss (J)	Terminal SOC (%)
Case 1	41.45	11,818	99.41
Case 2	42.13	12,334	99.18
Case 3	40.76	11,943	99.09
Case 4	40.08	11,986	99.21

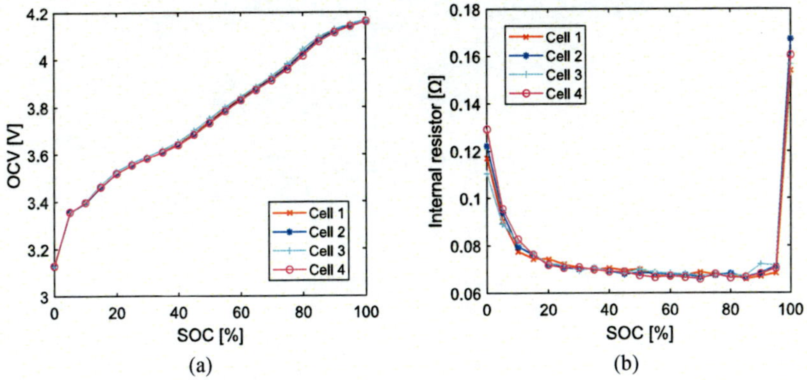

Fig. 10.6 The cells' **a** OCVs, **b** internal resistances

10.5 Experimental Results

In order to validate the effective of the designed charging control strategy, hardware-in-the-loop experiments are carried out on a battery pack consisting of four Panasonic NCR18650B battery cells connected in series, where the cells' capacities are identified as 3.058 Ah, 3.111 Ah, 3.024 Ah, 3.102 Ah, respectively. The cells' OCVs and internal resistances are illustrated in Figs. 10.6 (a) and (b), respectively. The maximum charging current that the charger can provide is 3 A. The maximum equalizing current of the equalizers is 0.3 A. The cells' maximum allowed charging current is 0.5 C-rate recommended by the battery's instruction manual, which means that this battery pack needs longer charging time than the simulated one with an upper bound of 3 C-rate, and the corresponding energy loss will also be reduced. Hence, the maximum energy loss of the battery pack is limited to 2% of its recharged energy in the experiment. Other parameters in the proposed battery pack charging control method are selected as the same in the above simulation.

Figure 10.7 illustrates the experimental test bench, which is composed of an Arduino Mega 2560 board and 4 self-developed controllable constant current source circuits. The Arduino Mega 2560 board samples the cells' signals and based on them computes the optimal currents that should be supplied by the charger and cell equalizers. Then, the self-developed controllable constant current source circuits

Fig. 10.7 Hardware-in-the-loop experimental platform

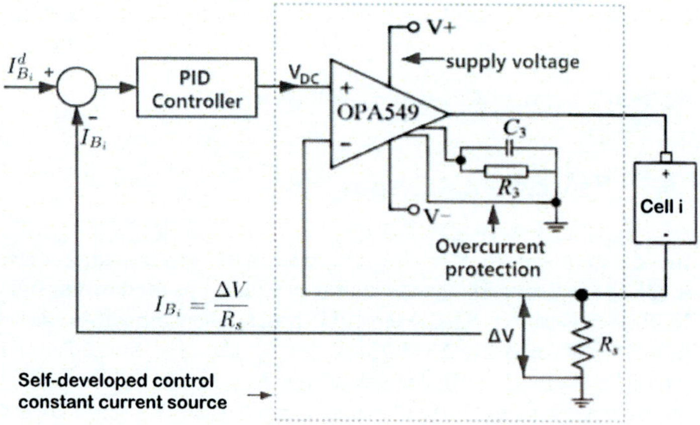

Fig. 10.8 Self-developed controllable constant current source

driven by the proportional-integral-derivative (PID) control algorithm provide the corresponding total charging current to the cells in the battery pack. The detailed description of the controllable constant current source circuit can be seen in Fig. 10.8, whose main component is a high-current operational amplifier OPA549.

The cells' initial SOC vector is set to $x(0) = [5\%, 0\%, 8\%, 10\%]^T$, and the desired SOC of the battery pack is 100%. The experimental results of the SOC and energy loss of the battery pack, the SOCs, currents, terminal voltages, and temperatures of the cells are shown in Fig. 10.9. The battery pack's SOC can be charged from

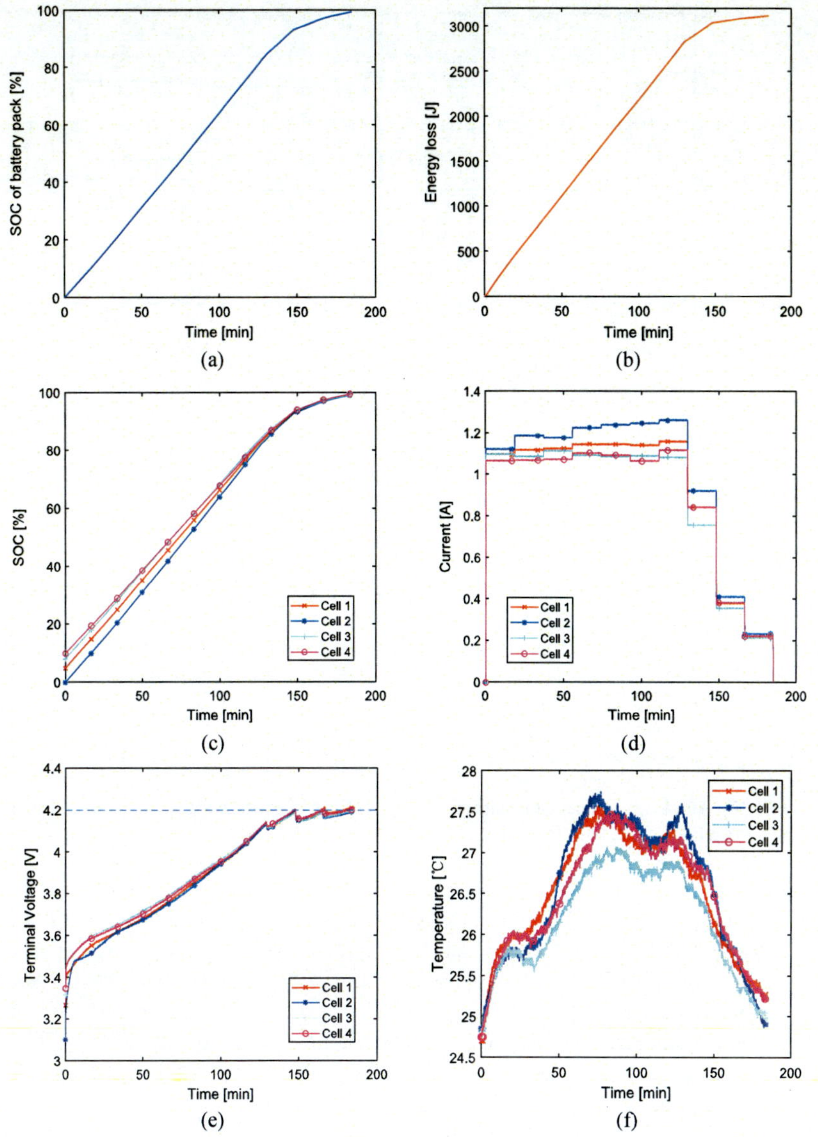

Fig. 10.9 Experimental results of **a** SOC and **b** energy loss of the battery pack, **c** SOCs, **d** currents, **e** terminal voltages, and **f** temperatures of the cells for $x(0) = [5\%, 0\%, 8\%, 10\%]^T$.

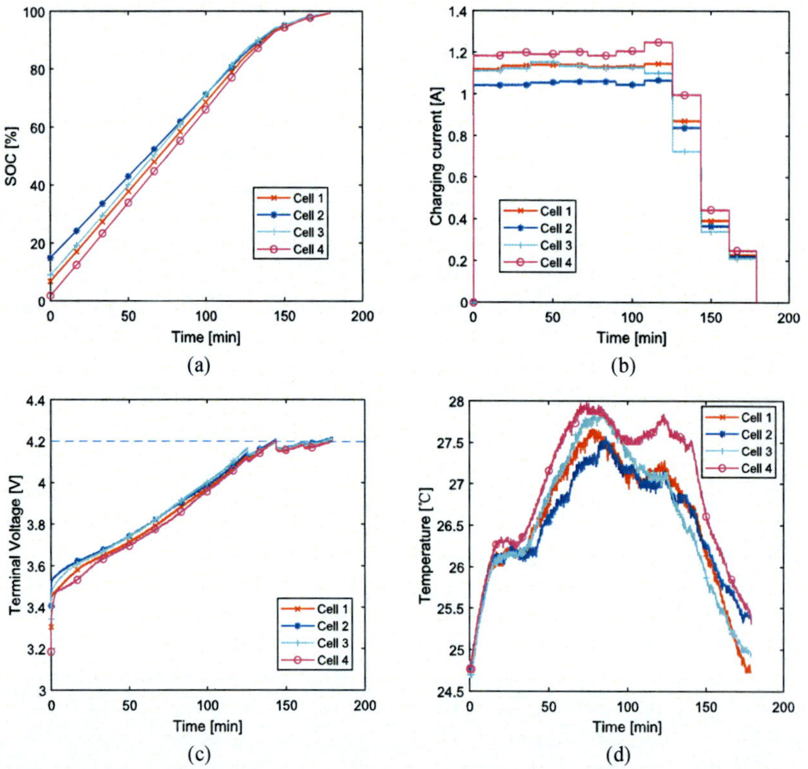

Fig. 10.10 Experimental results of cells' **a** SOCs, **b** currents, **c** terminal voltages, **d** temperatures for $x(0) = [7\%, 15\%, 9\%, 2\%]^T$.

0% to 99.39%, where the cells' SOCs, respectively, are 99.53%, 99.39%, 99.49%, and 99.82% after 185 min of charging. The battery pack's actual energy loss is 3, 122.2 J, which slightly exceeds its upper bound 3, 087.6 J. That is because of the battery model bias and the actual charging current derivation provided by the electrical circuits. The cells' maximum temperature can be about 27.75°C. These results show the effectiveness of the designed optimal fast charging method for the battery pack.

To further verify the effectiveness of the proposed battery pack charging method for different initial cells' SOC distribution of the battery pack, the experimental results shown in Fig. 10.10 for the battery pack with another initial SOC vector of $x(0) = [7\%, 15\%, 9\%, 2\%]^T$. After 179.5 min of charging, the battery pack's SOC can be charged from 2% to 99.39% with the cells' terminal SOC vector of $x(N) = [99.39\%, 99.62\%, 99.57\%, 99.53\%]^T$, while is energy loss is 3, 025 J, which is close to the pre-set limitation $J_{eM} = 3, 005$ J. The maximum temperature of the cells is

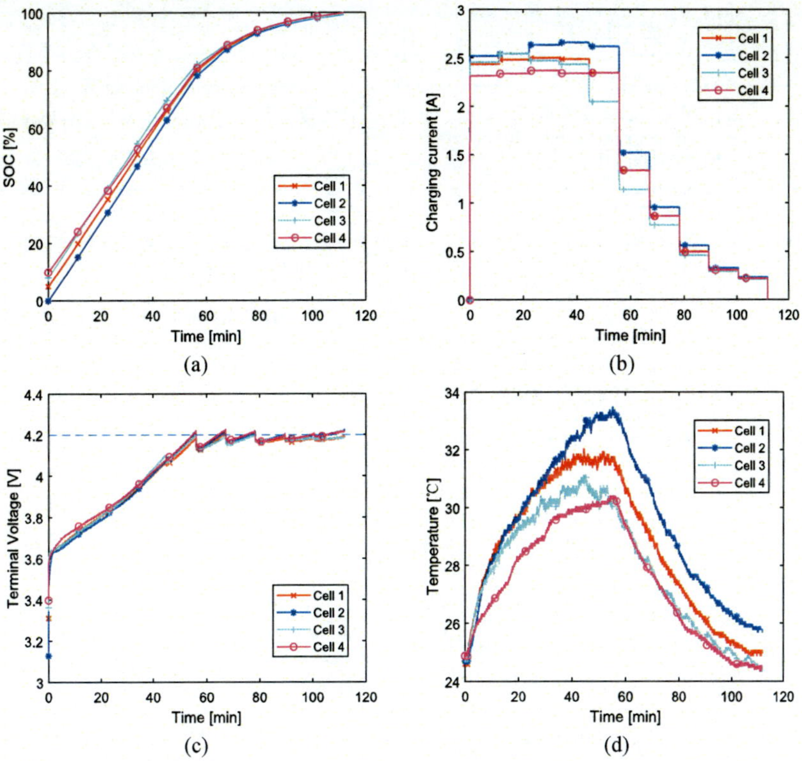

Fig. 10.11 Experimental results of cells' **a** SOCs, **b** currents, **c** terminal voltages, **d** temperatures for $x(0) = [5\%, 0\%, 8\%, 10\%]^T$.

about 27.99°C. It demonstrates that our designed charging control algorithm can work excellently for the battery pack with different cells' initial SOCs.

Finally, to verify the effectiveness of the proposed battery pack charging control strategy on high C-rate current charging, the maximum allowed charging current of the cells is set to 1 C-rate, and the maximum allowed energy loss of the battery pack is selected as 4% of its recharged energy. For the battery pack with the cells' initial SOCs of $x(0) = [5\%, 0\%, 8\%, 10\%]^T$, the charging experimental results are shown in Fig. 10.11 with the consumed charging time of 111.8 min. The battery pack's SOC can reach 99.33%, where the cells' SOCs are 99.48%, 99.33%, 99.43%, and 99.91% after charging, respectively. The energy loss of the battery pack is 6, 194 J. These results demonstrate the superior performance of the proposed battery pack charging control strategy. Note that the cells' maximum temperature in the battery pack is about 33.48°C, which is higher than the above results with 27.75°C for the energy loss of 3, 122.2 J. It shows that the battery pack's temperature rise can be restricted by limiting its energy loss, which is consistent with the above analysis.

References

1. Q. Ouyang, G. Xu, H. Fang, and Z. Wang, "Fast charging control for battery packs with combined optimization of charger and equalizers," *IEEE Transactions on Industrial Electronics*, vol. 68, no. 11, pp. 11076–11086, 2021.
2. J. Gallardo-Lozano, E. Romero-Cadaval, M. I. Milanes-Montero, and M. A. Guerrero-Martinez, "Battery equalizing active methods," *Journal of Power Sources*, vol. 246, pp. 934–949, 2014.
3. S. Jeong, J. Kwon, and B. Kwon, "High-efficiency bridgeless single-power-conversion battery charger for light electric vehicles," *IEEE Transactions on Industrial Electronics*, vol. 66, no. 1, pp. 215–222, 2019.
4. Q. Ouyang, W. Han, C. Zou, G. Xu, and Z. Wang, "Cell balancing control for lithium-ion battery packs: A hierarchical optimal approach," *IEEE Transactions on Industrial Informatics*, vol. 16, no. 8, pp. 5065–5075, 2020.
5. R. Marler and J. Arora, "Survey of multi-objective optimization methods for engineering," *Structural and Multidisciplinary Optimization*, vol. 26, no. 6, pp. 369–395, 2004.
6. S. Boyd, L. Vandenberghe, *Convex Optimization*, Cambridge University Press, New York, NY, USA, 2004.
7. Q. Ouyang, J. Chen, J. Zheng, and H. Fang, "Optimal multiobjective charging for lithium-ion battery packs: A hierarchical control approach," *IEEE Transactions on Industrial Informatics*, vol. 14, no. 9, pp. 4243–4253, 2018.

Chapter 11
The Future of Lithium-Ion Battery Charging Technologies

Lithium-ion batteries play an essential role in many applications stretching from electric vehicles to energy storage systems due to their advantages of higher energy density and cell voltage, as well as a longer life span and lower self-discharge rate. However, the high requirements of lithium batteries for charging conditions limit the widespread application of lithium batteries, particularly associated with battery damage due to over-charging, over-discharging, over-current, and other improper operations. Hence, scientific charging technology is necessary to optimize the performance and enhance the safety of the batteries. Considering the various approaches to lithium-ion battery charging in the above chapters, advanced model-based methods seem like the best bet for health-aware optimal battery charging. Various aspects of fast charging have been the subject of significant research in recent years, but there are still many knowledge gaps. In the future of lithium-ion battery charging technologies, three elements will be increasingly crucial: multi-objective optimization-based charging technologies, high efficient battery pack charging strategies, and wireless charging. To this end, this chapter outlines some subjects regarding future investigations concerning lithium-ion battery charging technology.

11.1 Multi-objective Optimization-Based Charging Technologies

Currently, the fast charging techniques are booming that can increase the charging speed. However, the blind pursuing the single objective of fast charging will inevitably bring about problems such as rapid battery capacity degradation. In practice, many other indicators, such as the charging time, charging energy loss, temperature risen, and capacity degradation, are important optimization objectives to be considered in charging control for lithium-ion batteries. Hence, the multi-objective

Q. Ouyang and J. Chen, *Advanced Model-Based Charging Control for Lithium-Ion Batteries*, https://doi.org/10.1007/978-981-19-7059-7_11

optimization-based charging control strategies shows their advantages, which can obtain the optimal current that can balance multiple objectives. They can achieve multiple objectives by self-adjusting the charging current based on the user specification and the battery's characteristics, which can result in a lower electricity bill, less energy loss, reducing the battery's capacity loss, and so on. The multi-objective optimization-based charging control issue can be transferred to a constrained optimization problem, and many intelligent methods can be used to solve it to get the optimal charging current.

11.2 High Efficient Battery Pack Charging Technologies

Lithium-ion battery packs rather than battery cells are often used in practical applications such as EVs, energy storage system, etc. Hence, the characteristics of the battery pack will be studied more thoroughly, and high efficient battery pack charging strategies are the future trend, which cannot only satisfy the considered multiple objectives but also achieve cell equalization. Note that battery pack charging control brings significant computational burden increase. Advanced artificial intelligence techniques are effective potential solutions to optimize charging for individual cells in a pack that contains multiple connected cells to provide better charging performance by avoiding unwanted limitations, overcharging, and local degradations. Multi-scale modeling will also be imperative to integrating individual cell and pack level design and control. It will facilitate the link between research on different scales and advancements in commercial systems.

11.3 Wireless Charging Technologies

With the rapid popularization of electronic products, the user's demand for charging technology is gradually developing toward intelligence and convenience. The subsequently proposed wireless charging technology can automatically and intelligently charge electronic devices without manual cable connections. Compared with wired charging techniques, the wireless charging strategies have certain advantages in adaptability and safety. The wireless charging technology has an anti-infection solid ability for the environment, and the harsh environment will not affect the charging operation of the wireless charging device. In addition, the wireless charging device has simple operation and low promotion cost since it is easier to promote into practical applications. As an emerging technology, wireless charging techniques have been commercialized in low-power devices such as mobile phones and computers. The applications of high-power devices for charging EVs have also become the focus of major scientific research institutions and commercial EV companies.